Concrete Leviathan

Concrete Leviathan

The Interstate Highway System and State Building in Modern America

TEAL ARCADI

The University of Chicago Press
Chicago and London

Publication of this book has been aided by the Meijer Foundation Publication Fund, which supports books of enduring interest in the disciplines of American political history, political science, and related areas.

The University of Chicago Press, Chicago 60637
The University of Chicago Press, Ltd., London

Published 2026
Printed in the United States of America

35 34 33 32 31 30 29 28 27 26 1 2 3 4 5

ISBN-13: 978-0-226-85139-6 (cloth)
ISBN-13: 978-0-226-85141-9 (paper)
ISBN-13: 978-0-226-85140-2 (ebook)
DOI: https://doi.org/10.7208/chicago/9780226851402.001.0001

Library of Congress Control Number: 2026012506

♾ This paper meets the requirements of ANSI/NISO Z39.48-1992 (Permanence of Paper).

Authorized Representative for EU General Product Safety Regulation (GPSR) queries: **Easy Access System Europe**—Mustamäe tee 50, 10621 Tallinn, Estonia, gpsr.requests@easproject.com
Any other queries: https://press.uchicago.edu/press/contact.html

For Andrea

Contents

INTRODUCTION

Road Work Ahead

"We know you will try to help us," wrote Nell McNeil to her congressman, Mike Mansfield of Montana, in the autumn of 1952. McNeil owned a Texaco gas station in a small town on the western edge of Yellowstone National Park. She had just learned of a new plan to replace the road alongside her business with a federal highway located on a new route some distance away. The community's business owners were "heartsick" from the news. The existing road supported their existence and expectations. "We all have had faith in our future here," McNeil explained. "We all worked hard. . . . We have reinvested every dollar we could spare to improve our place of business." The future that McNeil and her neighbors had imagined and constructed would be swept away if highway administrators went ahead with their plan. She "beg[ged] for economy in Washington, DC," hoping the federal government would choose to spend money maintaining the existing road rather than moving it to meet new ambitions directed by the Bureau of Public Roads (BPR), the agency that oversaw federal highway planning. "We protest vigorously that this new project be abandoned and that the old route be improved and maintained," she concluded. Her community would "lose everything" if the route change came to pass.[1]

Within a week, Mansfield reacted to McNeil's appeal, but not by sending her a reply. Nor did he forward her letter to local or state highway officials as he might have done a generation earlier. Instead, he sent her letter across Capitol Hill to Thomas MacDonald, commissioner of the BPR.[2] Asking MacDonald for information on the highway route McNeil opposed, Mansfield made clear that he had no knowledge of highway development plans in his own congressional district. Some two months later, MacDonald responded with his agency's position on route planning in Montana. "An economic study" favored the new highway plan that worried McNeil. Who conducted this study? Whose economic interests was it meant to serve? Who would control the process of turning the plan into pavement? MacDonald did not explain, noting merely that the study had garnered "the general approval of the official agencies involved." Mansfield was out of the loop, and McNeil was out of luck.[3]

Like Americans across the country in the postwar years, McNeil was caught in the final stages of a conceptual revolution that had gathered force since the early twentieth century. In that time, agency administrators like MacDonald had imagined a vital role for new federal highways in American political economy—and a central position for themselves in advancing that agenda, which culminated in the construction of the National System of Interstate and Defense Highways. The interstate highway system became the most expensive and expansive public works project in American history, expressing and manifesting officials' economic, social, and legal aspirations in a spatially embedded regime of government across the nation.

As McNeil's experience signaled, the leaders of that regime worked to insulate their authority from democratic input in the decades ahead. Officials brought the advantages of infrastructural connectivity to some areas and wreaked havoc on others by wiping them off the new map. Their beacon public works project, amounting to over 40,000 miles of asphalt, concrete, and steel modifications of the social landscape, made their worst impulses toward political domination and inequality starkly legible. In time, the infrastructure they built became a nation-spanning site in which citizens encountered the interstate highway system's uneven consequences and demanded a say in the foundational processes of government that had brought it into existence.[4] Too often, their demands went unanswered. But as this book explores, the history of the interstate highway system's problems with democracy, participation, and equity opens up powerful new possibilities for enhanced practices of democratic government in the modern United States.

*

Plans for the interstate highway system trace back to the 1930s, when BPR officials began to study how a network of federal highways connecting cities might boost national productivity. After the exploration of a series of possible route maps, the Federal-Aid Highway Act of 1944 formalized a 40,000-mile "National System of Interstate Highways" designed to link the nation's "principal metropolitan centers" and spur economic growth. But the 1944 act did not fund the infrastructure it planned out, although it mandated continued planning.[5] More than a decade passed before the Federal-Aid Highway Act of 1956 appropriated $27 billion to transform the plan into actual highways. A staggering sum, the total expenditure relied on $25 billion drawn from new federal excise taxes on gasoline, tires, and other automotive components, with states contributing the remaining $2 billion from their own revenue sources. The funds accumulated in a new entity called the Highway Trust Fund.[6]

The interstate highway system's funding soared past the New Deal's most consequential projects, still fresh in national memory. The 1933 National Industrial Recovery Act, for example, ran to $3.3 billion (much of which was spent by the Public Works Administration), and the Emergency Relief Appropriation Act of 1935 totaled $4.88 billion (funding the Works Progress Administration, among other agencies).[7] In today's money, the $27 billion appropriation is equivalent to over $1.6 trillion.[8] The deep pool of petroleum-derived capital shaped the highway project's financing in striking ways. Older transportation systems such as the transcontinental railroads saw government mask its priorities and support in associational relationships with private corporations and investors.[9] By contrast, federal agencies openly took charge of the interstate highway system's financing and development. Using earmarked tax revenue to fund construction, the federal government did not even raise capital from bonded debt issued by private lenders, as was widespread practice in public works projects at the time.[10]

The Highway Trust Fund, overflowing with capital and operating in plain sight, fell under the discretionary control of administrators in the BPR and, later, in the Federal Highway Administration. The arrangement enabled federal officials to build the interstate highway system how and where they pleased. The project—which quickly surpassed initial plans and eventually grew to over 46,000 miles in length—required vast quantities of sand, gravel, tar, steel, and other materials. The most important commodity was oil. Every time a motorist filled their gas tank or bought a tire, excise tax revenue flowed onward to interstate construction, demanding yet more oil to run heavy equipment, oil to mix into asphalt substrate, oil to power the cars, trucks, and buses whose petroleum dependence kept the cycle going. Gas station owners in more fortuitous locations than Nell McNeil's prospered when the interstates came their way. Road builders were just as dependent on petroleum for their livelihoods and did well no matter where the new highways led. Construction required an enormous workforce trained to clear and flatten routes, lay pavement, and build bridges in accordance with the interstate system's intricate technical specifications.[11]

As the pavement hardened, it fixed new patterns of political economy in place. Many urban workers—particularly those who were white and had access to affordable mortgages—became suburban commuters. Weekend vacations got easier, and long-haul truckers could drive goods from San Francisco to New York without hitting a stop light.[12] Acting on their earliest imaginations of a federal highway system in the 1930s, officials had established patterns of movement and existence that would define the United States from that point onward.[13] The durability with which the interstate highway system rewrote the nation's future was, from the perspective of its supporters, one of its greatest triumphs. "This new highway program will affect our entire economic and social structure," Robert Moses proclaimed in 1956, and "leave a permanent imprint on our communities and people."[14]

*

The ambitions that added up to the interstate highway system—to craft political economy via dominating physical infrastructure developed by robust administrative power—were characteristic of the "high modernist" era of statecraft that swept the world in the middle decades of the twentieth century. The political theorist James Scott defines high modernist ideology by its unshakable commitment to the "comprehensive planning of human settlement and production," and an equally rigid "self-confidence about scientific and technical progress." Some of the chief examples of mid-century high modernism included the construction of dams, communication and transportation systems, and even entire cities like Brasilia and New Delhi, all of which erased preceding social arrangements and imposed dramatic visions of centralized authority. The first steps in this high modernist movement arrived with the Great War. The unprecedented scale of fighting and the ensuing breadth of destruction revealed after the armistice prompted new national economic planning efforts in the United States, Europe, and the Soviet Union. A short time later, the even more profound devastation of World War II refracted the same planning urgencies and delivered high modernism in full in the war's aftermath. The year 1945 became a turning point in the high modernist interest in "the administrative ordering of nature and society." The ambition was to manage production and aggregate economic growth in the political and economic competition among nations that came to define the postwar era.[15]

In these currents of change, nothing guided imaginations of a new federal highway system in the United States so clearly as the rising interest in aggregate economic growth. New Deal efforts to comprehend and resolve the Great Depression had crystallized an innovative category of analysis called "the economy." Commonplace today, the economy was a novel concept at the time. As an organizing force, it gained further prominence when the nation worked to manage and expand productive capacity during World War II, and it then shaped the quest for prosperity after 1945. The economy's invention overturned understandings and expressions with which people made sense of society and its government. Nell McNeil, for example, had used an antiquated meaning when she asked government officials to exercise "economy" in her appeal to reconsider highway relocation. She meant something like thrift, asking the government to save money, forgo new development, and allow her business to continue. But the addition of the definite article made *the* economy something different: "an object in relation to which claims to a more egalitarian life could be measured and adjusted, and matters of common concern could be removed to the administration of experts," as the economic historian Timothy Mitchell puts it. If high modernism came with an interest in ordering nature and society, the invention of the economy offered new metrics and linguistic forms with which to justify government activity.[16]

By the time the interstate highway system became law, the economy was the organizing principle of policymaking. Like other aspects of high modernism, it required sophisticated, authority-laden, and well-resourced administrative bodies to turn economic imaginations into reality. Such administrative capacity had proliferated in the United States during the New Deal and World War II years. The growing collection of federal committees, boards, agencies, and other bureaucratic units drew new terminology, called the "administrative state" or even the "fourth branch of government," though its powers, at least in theory, were contained within the executive branch. Filled with emboldened officials and their rising authority, the administrative state drew a more foreboding nickname as well: "Leviathan." This evoked the biblical sea creature that had served as a symbol of central government ever since Thomas Hobbes borrowed its name for his treatise on the theme. Leviathan's association with mid-century federal power reflected not only the newfound depths of administrative governance but also the very essence of how the administrative state governed. It did not have a single purpose, instead assuming a role in any number of tasks that might previously have been left to local, state, or private decision-makers. Leviathan took physical shape with the interstate highway system.[17]

The mid-century administrative state grew from the seeds of progressive government efforts planted decades before. At its best, its officials aimed to put expert knowledge to work in pursuit of the public interest, checking the coercive and exploitive aims of private power and seeking to craft an equitable political economy for the nation's citizens.[18] This mission had frequently fallen short of its best ideals. Sometimes private interests captured the administrative state's regulatory mechanisms, and other times racism, sexism, classism, xenophobia, and other divisive forces warped the administrative state's actions in ways that created insiders and outsiders among the nation's inhabitants. These problems indicated design flaws. If elite administrators could concentrate their own power and act on narrow values and priorities, the administrative state contained suspiciously undemocratic tendencies. By the time the interstate highway project took shape, this realization had prompted a generation of "sympathetic critics"—supporters of the ideals of the administrative state who took issue with its frequent shortcomings—to warn that Leviathan's rising power came with risks, at least from the perspective of those seeking a fair social contract and an equitable national landscape of opportunity.[19]

What kinds of values and priorities would shape the interstate highway system? National economic growth was the guiding logic—a logic that contained a series of nested choices about where and how to build the new highways, who would benefit from them, and who would be burdened by their imposition. The economist Simon Kuznets, one of the first to conceptualize the economy in the 1930s, had issued a dire warning: The "welfare of a nation," he told Congress in 1934, could "scarcely be inferred from a measurement of national income," tantalizing as a soaring gross national product graph line might be.[20] Even if the nation's overall wealth increased, there was nothing in his modeling to suggest this meant that every community would come out a winner. This problem would be put to the test in dramatically physical terms when the interstate highway project etched the era's new economic ideas into the nation's geography.

*

When President Dwight D. Eisenhower authorized the interstate highway system's construction, he proclaimed it to be "the greatest public-works program in the history of the world." Bringing things down to earth, *The New York Times* celebrated the "tangible . . . benefits" the interstate system would bring to all Americans.[21] But only a decade later, a very different article in the *Times* captured changing sentiments when it lambasted the political establishment's heady embrace of highway development. Congress had "bow[ed] to the bulldozer" and prioritized pavement over "parks, historic sites and wildlife areas." Congress, according to the *Times*, had made clear its alliance with roadbuilding interests and, more conceptually, with the interstate system's long-standing promises of economic growth. The paper now claimed that Congress was not serving the interests of "local conservationists, citizen groups and poor residents," who drew attention to the interstate system's downsides with mounting urgency. The nation's lawmakers, the *Times* concluded, would be "memorialized by endless miles of a highway blight that has already terribly scarred the face of this once-lovely land."[22] In 1956, optimistic Americans could imagine a better nation yet to be built. By the end of the 1960s, the "nation's faith" in the interstate system, now almost complete, had been "painfully shattered" as maps and blueprints became asphalt and concrete.[23]

The devastation the interstate highway project brought was stunning. To make way for the system, the BPR displaced more than one million Americans, leveling neighborhoods in nearly every city the interstates touched.[24] A disproportionate amount of this destruction targeted poor, nonwhite areas, especially those with majority Black populations. Racist city officials used the process of interstate highway construction as a brutal—and federally funded—tool of so-called "slum clearance," destroying unwanted communities of color under the duplicitous banner of "urban renewal." Of course, the destruction of those communities did not always mean that they were cleared of residents. Often, destruction amounted to forcing communities to live in the shadow of urban overpasses, where they experienced a downward cycle of disinvestment and poverty in increasingly undesirable neighborhoods. The new highways spurred "white flight," with white homeowners and businesses moving out of cities into newly interstate-accessible suburbs that remained divided along race and class lines. After *Brown v. Board of Education* ended formal segregation in 1954, the arrival of interstate highway funding two years later offered a means of eviscerating Black spaces, privileging white ones, and building impassable multilane barriers between the two.[25]

These aspects of interstate highway development were immediately apparent. As soon as bulldozers and paving machines roared to life, "freeway fighters" in cities across the nation organized grassroots protests, marching in the streets to stop the destruction. There were some notable successes. As early as 1959, protestors in San Francisco made use of unusual provisions in California state law to force a review process that succeeded in altering an interstate's route. A few years later, Jane Jacobs and her allies helped secure another victory when they saved Washington Square Park from destruction.[26] By and large, though, freeway fighters failed to stop the relentless capital appropriations of the Highway Trust Fund and the pavement it churned out, especially when fighting racism was the basis for protest. In 1967, for example, a Tennessee community group filed suit to protect North Nashville—a historically Black area—from the construction of Interstate 40. *Nashville I-40 Steering Committee v. Ellington*, alleging infringements of due process and equal protection, became the first lawsuit over interstate highway construction to reach the Supreme Court—which promptly declined to hear the case, letting stand a lower court's conclusion that there was no proof of discriminatory intent. A few years later, the Supreme Court agreed to hear a different case, *Citizens to Preserve Overton Park, Inc. v. Volpe*. Dealing with the more politically popular goal of parkland protection, *Overton Park* established a precedent for judicial review of administrative action—a win, more or less, for the freeway fighters in the case. But the outcome did nothing to alter the form of administrative agencies and make them more inclusive of and accountable to communities. Instead, the case's result just pressed administrators to scheme more carefully when they targeted land for development.[27]

If freeway fighters could not avert interstate highway development, they succeeded in calling attention to the project's downsides—and to the democratic limitations of the administrative framework undergirding the interstate highway project. The concrete leviathan that came to embody the American state in the middle decades of the twentieth century made abstract tensions between administrative authority and participatory democracy into physical problems that could be evaluated in the built environment. The interstate system's routes were material manifestations of fault lines running through the very foundations of modern American liberalism. Plaintiffs may have gone to court to challenge the immediate perils of infrastructural state building—the destruction of a neighborhood or park. Yet as they imagined alternatives in their communities and cohered demands in courtrooms, questions surfaced regarding the societal roles of citizens, politicians, administrators, and judges—questions, that is, concerning the foundational elements of the nation's government.[28]

*

Infrastructure projects rarely impact all equally. The priorities they reflect and the inequalities they inscribe make them ideal subjects for evaluating government as it uses them to alter society.[29] Theorists and social scientists have come to recognize this over the past several decades. An important analytical starting point was sociologist Michael Mann's investigation of "infrastructural power" in his germinal 1986 work "The Autonomous Power of the State: Its Origins, Mechanisms, and Results." In infrastructure, Mann saw "the state's unique ability to provide a *territorially centralized* form of organization" (Mann's emphasis). The topic remained somewhat conceptual; the power Mann referred to amounted to the state's capacity to "penetrate civil society, and to implement logistically political decisions throughout the realm."[30] Two decades later, infrastructure as a conscious object of analysis had gained little traction among historians, although William Novak proved a key exception when, aiming to ground Mann's theoretical observations, he called for expanding our "understanding of infrastructural power" by historicizing "the actual power of state policies in action to have real effects on populations."[31]

More than another decade passed before historians began paying attention in earnest to the payoffs infrastructural analysis can bring. By 2023, the effort was in full swing. Journal articles appeared taking stock of the "infrastructural turn in historical scholarship" and even arguing that infrastructural analysis should be applied anachronistically to the distant past, in investigations of power dynamics in societies that did not even have infrastructure—or the concept thereof—as we know it today.[32] It was not a coincidence that this "infrastructural turn" arrived around the same time that Joe Biden's presidential campaign called attention to the "crumbling" condition of American infrastructure systems and pledged to "build back better." Insisting that "infrastructure investments" would "mitigate socio-economic disparities, advance racial equity, and promote affordable access to opportunity," the Biden administration underscored infrastructure's social significance.[33] What captured even more attention was the administration's admission that there were downsides to be avoided, too, in the effort to improve the nation's infrastructure. Marking a distinct break with the past, key federal administrators noted infrastructure's propensity to produce uneven benefits and burdens. "There is racism physically built into some of our highways," Secretary of Transportation Pete Buttigieg told the press soon after taking office.[34]

With the glaring examples of highways and racism, Buttigieg reminded the public that infrastructure is a means of inscribing priorities in the social landscape. In doing so, he indicated a concise, if abstract, definition of infrastructure that this book follows: Infrastructure is a site where "power is encoded," as the architectural historian Reinhold Martin once theorized.[35] Infrastructure, by its nature, is also a site that connects past, present, and future; for these reasons, analyzing how officials construct infrastructure systems and use them to govern across space and through time is a critical step in the production of alternate forms of government moving forward.

*

Studying infrastructure shines a light on the form and function of the government that creates it. The fractured landscape that resulted from the interstate highway system's development emerged, ironically, in the context of great reckonings with the inequalities of American economic life: born of Franklin D. Roosevelt's New Deal, constructed during Lyndon B. Johnson's Great Society, and contested amid the rise of the New Left. The "modern Republicanism" that shaped Eisenhower's support for the 1956 highway act, too, was interested in resolving some aspects of disparity within society. Seen one way, this makes the harms of highway development and the chronology of failed freeway revolts all the more disappointing. But it also makes them more revealing. Even the Civil Rights Act, the era's foremost revision of the American social contract, had little impact on the nation's defining public works project.[36]

The postwar decades, marked by new recognitions of inequality and important expansions of rights and opportunities, were also marked by consistent elite resistance to citizen power.[37] Administrative agencies involved in everything from business development to housing to poverty relief took steps to curtail all but the most token community participation in decision-making.[38] Where community input in infrastructure projects did occur, it was a product of race and class privilege, not democratic design. Reformers realized all of this. But broader efforts toward participatory democracy, organized around the concept of "citizen advocacy," backfired at precisely the same time—the 1960s and early 1970s—when community input in the interstate highway project might have averted some of its worst harms.[39] It was in these years that public figures like Rachel Carson, Jane Jacobs, and Ralph Nader came to prominence. Making claims about authoritarian control exercised by administrative agencies, they levied significant critiques of administrative action.[40] They recognized a simple truth: Too often, the unwieldiness of administrative governance had, in Nader's words, done "violence to equal protection for citizens," often because of a mix of undue influence by private interests and flawed institutional design.[41] But the movement these figures inspired struggled to formulate an actionable plan for more accountable and participatory government.

The only remedy that emerged was a heightened emphasis on lawsuits brought by individuals or interest groups against government agencies. This "citizen advocacy" approach bore similarities to the new civil rights law framework. Sometimes citizen advocates pursued an equitable social distribution of costs. For example, the plaintiffs in *Nashville I-40 Steering Committee v. Ellington* hoped to protect Black communities from targeted destruction by interstate highway planners. The refusal of the federal courts to recognize any evidence of racist agency action, however, indicated a significant problem with lawsuit-based agency reform efforts. Another problem was that doubling down on citizen advocacy gummed up the work of government. Anyone with a grievance could delay or halt agency activity, leading to still more administrative problems and public frustrations—indeed, the obstruction of genuine public goods. In this way, accusations of administrative ineptitude coming from both the left and the right became something like a self-fulfilling prophecy. The result was to weaken, rather than enhance, the very governmental capacities that were failing to equitably protect public interests and provide public goods. While Nader and his fellow travelers sought better government, they undermined already flawed and vulnerable institutions. Exaggerating the inefficiency and illiberalism of the administrative state, critiques from the right—typified by rising public intellectuals and politicians like Milton Friedman and Ronald Reagan—called with increasing fervor for its total abolition.[42]

The racist pattern of interstate highway construction wrought immeasurable damage, revealing in the process an expansive set of challenges relating to democracy and inequality in the work of state building. And the problem of racist infrastructure remains largely unresolved. The civil rights framework that has existed since 1965 cannot account for and counter infrastructural harms at the level of the community—a significant deficiency, if government targets entire communities for destruction on the basis of race or other discriminatory factors.[43] Another fundamental and unresolved issue is the democratic potential of the administrative state. Sympathetic critics warned as early as the New Deal that administrative authority could become untethered from community input and need; citizen advocates in the ensuing decades could not provide a solution to the problem. Until that foundational issue is resolved, policymaking and infrastructural development of the kind typified by the interstate highway system will remain painfully skewed.

*

By the time historians made the "infrastructural turn," communities across the nation were also encountering new infrastructural possibilities. Like other examples of modern American infrastructure dating back to the mid-twentieth-century development boom, the interstate highway system began to reach the end of its planned physical life by the twenty-first century. Mile after mile of pavement became dilapidated beyond repair. For communities that had long suffered from infrastructural burdens, this presented an opportunity. In Syracuse, New York, for example, activists pushed the city—successfully, in the end—to remove a deteriorating stretch of Interstate 81 whose construction in the 1960s had destroyed the city's historically Black 15th Ward. Demolishing that portion of I-81 and rerouting it through Syracuse's outskirts—where wealthier, whiter suburbs decried its presence—looked like poetic justice. But significant questions about what would take the highway's place in the city center remained. Could a vibrant, mixed-income, and mixed-use community, as existed before I-81's imposition, be recreated? Or would the city prioritize high-end development and gentrification? Would the city's residents have any say in the matter, either way?

Issues like these place Syracuse in the middle of national policy debates about how to undo the racist and inequitable choices that guided the interstate system—and how to avoid similar problems moving forward, with respect not only to highways but to all large-scale infrastructure systems. The infrastructural issues in Syracuse proved to be a catalyst for novel ideas about legal and governmental reform. A promising starting place has been offered by Deborah Archer, the legal scholar and American Civil Liberties Union president whose work is at the vanguard of reimagining infrastructure and community equity. She proposes a federal requirement for "racial equity impact studies" in infrastructure planning and development. The current political climate makes this is a decidedly aspirational idea at the federal level. But there are significant state and municipal examples already in effect. These subnational models take cues from the environmental impact statements required in all federally funded actions by the National Environmental Policy Act of 1969 (NEPA). Scaled up to the federal level, racial equity impact studies would employ sophisticated historical, legal, and social science methods to evaluate community risks present in any infrastructure project. Archer's proposal recognizes, however, that this approach may come with a fatal flaw. Environmental impact statements are notoriously unenforceable. They require administrators to gather information but do not require that information to guide action. For community equity evaluations to have any significance, legal frameworks and the administrative state itself will have to change.[44]

On this front, one possibility—a lofty one—is the creation of a new agency tasked with upholding community equity and carrying out the challenging social and technical work of evaluating risks and evenly distributing the social benefits and burdens of infrastructure projects. Points of inspiration for such an agency exist. A key one is the Consumer Financial Protection Bureau (CFPB), legislated in the wake of the 2008 Great Recession. The CFPB's founding mission was to serve the public interest, enforcing rules of fairness on private financial institutions. By intention, its design made it a site of grassroots protest and knowledge sharing. Bureaucratic innovations like publicly accessible databases enabled citizens to report exploitative private financial practices. And expert administrators—upholding their duty to the interests of the citizens—could take action and pursue enforcement. The CFPB's planners also nested its funding within the Federal Reserve System, hoping this would shield it from capture by private interests whose behavior it is meant to check—and their political allies.[45]

That framework held strong over the first decade and a half of the CFPB's existence, with the agency rebuffing attacks by interests opposed to its democratization of power and its ability to regulate a wide range of private economic actors. The most serious threat yet faced by the agency came from the second Trump administration, which moved quickly to use executive power to shutter the CFPB early in 2025. The US District Court in Washington blocked that action but likely did not resolve the matter. Executive branch officials have continued to chip away at the agency's activities, and by the summer of 2025 conservative congressional leaders succeeded in cutting the CFPB's funding by half. Those efforts followed deeper attempts to undermine the administrative state's entire structure.[46] Ironically, less than a year before the second Trump administration began, the Supreme Court had upheld the constitutionality of the CFPB's unusual funding structure in *CFPB v. Community Financial Services Association of America, Ltd.*—a ruling that gave the CFPB's allies hope the agency would continue to thwart its adversaries. But that decision came alongside a slew of others that dramatically altered the landscape of administrative law and governance. Of particular significance was *Loper Bright Enterprises v. Raimondo*, which saw the court shift power away from expert federal administrators and into the hands of judges.[47] As legal commentators noted, this was a milestone in conservative jurists' ongoing effort to write themselves "a blank check to void any regulations" they wish. While it remains to be seen how the *Loper Bright* ruling will play out in the years ahead, it is the most direct assault yet on the essential structure of the American administrative state.[48]

Some of that administrative state's most authoritarian elements were on display throughout the long history of the interstate highway system, stretching back nearly a century. Reform is overdue. But decisions like *Loper Bright* hardly provide a constructive blueprint for new agencies empowered by Congress and managed democratically for the public good. In an era when the very existence of effective public administration is under dire threat, the challenge is to counter with a successful effort to build more equitable institutions from the ground up. One can envision, for example, an agency sharing the CFPB's commitment to democratizing the administrative state and crafting a fairer society. A "Community Equity Protection Bureau" could enforce new laws around infrastructure planning and development practices, serving as a sharing point for citizens encountering problems and a clearinghouse for administrators pursuing infrastructural justice. Such an agency would require careful design to resist anti-democratic factions, with thick insulation around its financing and institutional structure. Like the CFPB, its funding and authority might bear a disconcerting resemblance to the Highway Trust Fund, whose unchecked capital stream and disconnection from oversight made it the engine of the interstate highway system's worst consequences. It is deeply ironic that the best solution to undoing and avoiding infrastructural harms caused by flawed administrative practices may be the creation of strong new administrative apparatuses of a different variety.

*

FIGURE 1. International Harvester bulldozers ready for service, 1944. Wisconsin Historical Society, WHI-63853.

A considered effort to devise an administrative apparatus for equitable infrastructural development is the complicated capstone to the history of the interstate highway system—and the jumping-off point with which this book ends. Before that point, the chapters ahead explore how the interstate highway system embodied the aspirations of the officials who made it a reality. In total, the book offers a historical narrative of an infrastructure system that continues to express and manifest fraught priorities and practices of American government, from the New Deal to the present. Across six chapters, the book explores officials hatching plans, congressional leaders arguing over the details and establishing unparalleled infrastructural powers, administrators producing a robust culture of dominance to guide their work, critics catching sight of the undemocratic regime taking hold, plaintiffs trying to fight back in court, and, finally, a small city in upstate New York at the epicenter of twenty-first-century debates about how to construct democratic government and infrastructure systems to match.

In 1952, Nell McNeil concluded her letter to her congressman with a simple question: "May we have your support, please?" Her entreaty, imploring Congressman Mansfield to halt development of a new highway that threatened her livelihood, reached the deepest halls of power, finding its way to the desk of BPR Commissioner MacDonald. The only answer McNeil ever received came in the form of a bulldozer a short time later. The question today is how to reimagine and rebuild an administrative state, one that responds to citizens and brings them into the work of building an equitable nation through truly democratic government.[49]

PART ONE

Constructing

1

Gasoline Dreams

On a foggy December afternoon in 1936, a group of highway planning officials gathered in a ballroom at the St. Francis Hotel in San Francisco. They were there for the annual meeting of the American Association of State Highway Officials (AASHO), and Philip La Follette, governor of Wisconsin, was due to give a headlining address. The governor, however, canceled at the last minute. Looking to fill the time, the event's organizers asked H. S. Fairbank, chief of the federal Bureau of Public Roads' Information Division, to take the stage instead. He seized his chance, speaking at length from a twenty-four-page document he had brought with him to the convention. By the time he concluded, he had laid out an earth-shattering policy proposal that would alter the course of national development from that moment onward. While La Follette planned to talk about how states and municipalities could chip away at funding local public works projects, Fairbank offered the audience an entirely different scale of infrastructural investment for consideration. He asked the highway officials around him to imagine the construction of a novel federal highway system stretching from coast to coast, spurring economic growth across the national landscape. It was a watershed moment: From that point forward, highway officials charted a path that led to the Federal-Aid Highway Act of 1956 and the National System of Interstate and Defense Highways, both of which continue to shape American political economy today.[1]

Fairbank's unexpected speech could not have been timed better. A month earlier, American voters had sent Franklin Roosevelt back to the White House in the greatest landslide presidential election in modern American history. Securing over 60 percent of the popular vote, Roosevelt's resounding victory underscored the persistence of the nation's economic struggles and the continued faith that his administration could provide solutions.[2] But as immiseration and uncertainty continued, reform efforts that had marked the early years of the New Deal faced a political onslaught from conservatives and liberals alike. Leading economists—John Maynard Keynes foremost among them—and politicians began to reorient from direct relief measures to more abstracted theories of macroeconomic management aimed not at putting wages in peoples' pockets, but at summoning a rising macroeconomic tide that would, they claimed, lift all Americans with it.[3] The argument Fairbank made for a national highway system offered a concrete plan to put just such a model of aggregate economic growth into action.

Launching his plan for a national highway system, Fairbank characterized existing American highways as "inadequate." It was time to make highway development a national project, reimagining it to serve the "national economy" rather than the "narrow economic grounds" of local needs.[4] If that conception appears obvious today, it broke sharply from long-standing approaches to highway policy. The existing framework stretched back to 1893, when the federal government established the Office of Road Inquiry (ORI). The ORI's chief task was to map rural areas, often in collaboration with state agricultural colleges and their experiment stations, in order to plan rural roads and "farm-to-market" pathways. When the ORI became the Office of Public Roads and Rural Engineering, in 1905, officials maintained those emphases. From these efforts came the first direct federal funding for roadworks, in 1912, with a $500,000 grant for rural postal route improvements. This was characteristic of early federal roadworks: a focus on rural communities, and limited funding for the postal system.[5]

A more pivotal infrastructural intervention came in 1916. That year's "Act to provide that the United States shall aid the States in the construction of rural post roads, and for other purposes," also known as the Good Roads Act, signaled a new and more far-ranging federal interest in improving overland transportation routes. As the act's formal name implied, it channeled federal funds to subnational units of government. It required states to match federal funds, ensuring that state governments remained significantly responsible for their highways. While the act's $75 million federal contribution was a dramatic increase from the 1912 grant, the focus remained on postal delivery routes in rural communities. Yet the 1916 act set changes in motion: States' federal-aid eligibility became dependent on the establishment of state highway departments, setting the stage for coordinated highway planning and construction while somewhat masking the federal government's society-shaping influence.[6]

The conditions attached to federal funding turned state highway departments into satellites of federal authority, even as state highway administrators, often working closely with county or municipal officials, became the most visible directors of highway development from 1916 onward.[7] With this arrangement in place, the Federal Highway Act of 1921 appropriated another $75 million on the same terms as the previous legislation. The funds could be spent on up to 7 percent of a given state's rural roads, and states had discretion over which roads to fund. By the end of the 1920s, over 200,000 miles of improved rural roads had been built with federal aid in accordance with federal guidelines. This aid had, it appeared, "pull[ed] the farmer out of the mud" and funded the construction of important routes of regional economic life.[8] Nevertheless, rural development continued to define federal aid for road work: In 1932, federal legislation removed the 7 percent ceiling on rural mileage and allowed a 1 percent annual increase funded by continuing federal aid.[9]

Four years later, Fairbank, onstage in San Francisco, summed up these preceding efforts as "rough hewing." It was time, he said, to go a step further and begin to do the work of a "sculptor" aiming "well-directed blows" at a previously "formless mass." He asked his colleagues to carve automotive connectivity with "calculated" precision. And he made his ambitions plain. The nation needed a "highway *system*," he emphasized. It was no longer productive for roadworks to be controlled by "short-termed, relatively untrained local officials, each within his own narrow territorial limits, and largely without conscious continuing plan or coordination." The limited results of that approach were clear: State and local officials had failed to construct a system of highways capable of supporting "the economic and social structure of the nation."[10]

Fairbank's presentation signaled a shift in federal thinking. The goal was no longer to link discrete origins and destinations, or discrete producers and consumers—the "farm-to-market" roads of yore. Instead, Fairbank asked his colleagues to imagine a network of interconnections greater than the sum of its parts. Highway development was already on the upswing, to be sure, but even the famed Lincoln Highway, running from San Francisco to New York City, looked to Fairbank's rising sensibility like little more than a gussied-up stagecoach trail—ably connecting distant termini but hardly cohering the national geography. Fairbank recognized that "intelligent highway planning" could achieve far more.[11]

Orienting highway planning to the growth of the national economy rather than to local economic life was the first bombshell Fairbank dropped. The second was his argument that this required a highway *system*, a carefully planned network sculpted across the national landscape. He had a third bombshell in store, too: "Principal effort" in constructing such a system should be directed toward establishing "intercity" connections. Highway development would no longer be dedicated "to get[ting] the farmer out of the mud." Instead, it would turn to the cities. The BPR, Fairbank reported, had

FIGURE 2. The Executive Committee of the American Association of State Highway Officials at the organization's annual convention in San Francisco, 1936. Thomas MacDonald, Commissioner of the Federal Bureau of Public Roads, is seated fourth from right. © 1936, California Department of Transportation, all rights reserved.

determined that "a relatively small part" of the nation's total highway mileage "would serve the needs of the greater portion" of the nation if it provided "direct connections between the larger and more important cities."[12] Though nearly half the population still lived in rural areas, that number was slipping, and the nation's future looked increasingly urban. In sum, Fairbank called for the creation of a federally planned, urban-oriented American future riding atop nation-spanning bands of asphalt and concrete.

*

Fairbank's ideas were at the vanguard of a growing realization that a highway system could play a role in national economic growth, particularly a pattern of growth organized around cities. His attention to facilitating aggregate productivity fit the trends of emergent economic thought, too. Like other AASHO attendees, Fairbank worked in the flow of political and economic changes that had altered the course of the New Deal and American government. During Roosevelt's first term, on-the-ground relief efforts had provided direct aid and employment to millions of Americans in beacon initiatives like the Civilian Conservation Corps (CCC) and the Works Progress Administration (WPA). These agencies put people to work on environmental and public works projects, putting paychecks in pockets and getting capital circulating through the nation once more. But more abstracted approaches to government management of the economy were ascendant. As decision-makers set about remodeling American political economy, they hewed closely to the proposals that flowed from Keynes's revolutionary study, *The General Theory of Employment, Interest and Money*, published in February 1936.

Keynes's work—amounting to five hundred pages of data, analysis, and interpretation—offered a blazing critique of classical economics and a contrasting method for structuring economic growth. It quickly became a guiding light for American liberalism. Keynes's work maintained a commitment to government activity and social welfare. But the practices he helped devise and popularize involved far less direct redistributions of capital than in the New Deal's early days. The goal was no longer to use government to provide direct employment and public goods and services; "the economic system as a whole," as Keynes put it, became the target of intervention.[13] Macroeconomic mechanisms, supposedly, would stabilize and expand the private economy by ramping up demand-driven production, distribution, and consumption.[14] "The end of reform," as historians have termed it, had arrived. The hope for some was that central government could continue to produce prosperity, but without engaging so closely in lives and institutions as it had a few years earlier.[15]

Keynes's "economic system" became known as simply "the economy," a new term and category of analysis that encompassed complex aggregations of employment, investment, production, distribution, consumption, and synthetic averages of rates, prices, and wages.[16] The referent for these new measures was the nation, a politically definable area within which aggregates and averages could be measured. Novel national accounting practices like gross national product (GNP), meanwhile, produced a new language of government that allowed officials to quantify progress (or lack thereof) and define national territory as an economic area subject to management by experts.[17] All of this enabled new ways to imagine growth. Previously, economists had theorized growth as a matter of physical expansion predicated on harnessing resources, often through colonization. But in a modern world filled with nation-states and supposedly rigid demarcations of productive space, "internal intensification" of demand, production, distribution, and consumption—the economy's base factors—became the necessary substitute.[18]

The moment was primed for highway administrators to show how their infrastructure could, if properly constructed, contribute to this agenda. Once reimagined as a networked system, highways, according to Fairbank, would serve as the "new economic and social forces" of American productivity. Binding the economy and its territory together, modern highways offered "a nation on wheels" a means of speeding internal exchange and marshalling control of space and time.[19]

The nation was, indeed, one that ran on wheels. Automobile ownership rose from just under two hundred thousand vehicles in 1908 to over eight million in 1920. By the mid-1930s, 90 percent of American households owned a car.[20] And the rough-hewn roadways were not up to the influx. *Fortune* magazine called attention to the problems in 1936, decrying the "patchwork" of roads that required motorists to navigate "without rhyme or reason from dirt to macadam and concrete; from two-lane to three-lane to four-, five-, and six-lane breadth, from well-calculated curves to slithering turns." The BPR conceded the problem, concluding in 1937 that "existing highways are not equal to the burden now placed upon them by the modern motor vehicle."[21] Officials and observers had known as much for decades.

*

The need for basic improvements had long been obvious, even if a new economic regime was now required to inspire infrastructural investment. As far back as July 1919, a dramatic military exercise had heightened federal officials' awareness of highway deficiencies. In the aftermath of the Great War, the military decided to test two closely connected potential problems. One was the possibility that its motor vehicles were not up to the rapidly changing tasks of modern warfare. The other was that the nation's roadways were poorly prepared to facilitate citizen evacuation and military maneuvers, should fighting ever reach American shores. Both issues would be put to the test as the US Army's First Transcontinental Motor Convoy set out for California from Washington, DC. The convoy was composed of twenty-four officers, 258 enlisted soldiers, fifteen War Department staff officers, several dozen military vehicles, and assorted civilian cars and motorcycles. A fifteen-piece band provided by the Goodyear Tire and Rubber Company added to the spectacle—and indicated there were economic interests attached to the exercise, too, as the Army's road trip suggested the rising possibilities of civilian automobility.[22]

Along for the ride was a twenty-eight-year-old lieutenant colonel named Dwight D. Eisenhower. He later recalled that the exercise was prompted by the need "to search out the military capabilities of automobile and truck," and led by "enthusiasts for a transcontinental highway."[23] The convoy drew crowds and drummed up highway interest across the country. At one typical stopping point, a reporter noted that "practically the entire population" of East Palestine, Ohio, greeted the vehicles and men. That evening, the Army rewarded the town with a demonstration of a three-thousand-candlepower searchlight, shining a beam three miles high.[24] When the convoy arrived in

FIGURE 3. A day in the life of the Army's First Transcontinental Motor Convoy, 1919. Courtesy of the Dwight D. Eisenhower Presidential Library and Museum.

Chicago a week later, "several automobile delegations" welcomed it. A reporter there noted that Chicago—famed gateway to the West—was until then the westernmost point reached by a military "truck train."[25]

While the convoy spurred interest in highway development, its slow progress worried officials. It reached California after sixty long days that showcased just how bad the nation's roads were, especially for the era's less-than-formidable automotive technology. The main problems included mud slicks that engulfed the era's smooth, narrow tires; steep grades that overwhelmed engine torque; and bridges too weak to support even the smaller vehicles. Breakdowns were constant, regardless of road condition: The convoy averaged six miles per hour, traveling ninety miles on its best day and only four on its slowest. "Under the circumstances," Eisenhower wrote, the exercise was "a genuine adventure."[26] For the War Department, the operation demonstrated that "a comprehensive system of national highways, including transcontinental or through routes east and west, north and south, is real and urgent as a commercial asset to further colonize and develop the sparsely settled sections of the country, and finally as a defensive military necessity."[27]

Those concerns underscored the limitations of the state-level practices that had produced uncoordinated and poorly maintained roads. Meanwhile, the financial approach to road policy was changing. Road work was shifting from "a labor-intensive enterprise to a capital-intensive one," according to historian Christopher Wells. Local "labor levies"—mandatory workdays for local residents—had often accounted for early road work, supplemented by a growing array of state-issued bonds, property taxes, license and registration fees, and even poll taxes. The Great War upended these traditional sources of labor and capital, spiking construction costs at precisely the moment when automobiles were taking America by storm.[28] What enabled the turn to automobility was the same factor that might solve the growing road cost problem. By the time the 1919 convoy caused military and economic planners to consider federal highway development, a seemingly inexhaustible supply of cheap oil had begun to drive the postwar economy.

Between 1919 and 1929, gasoline distribution nationwide rose from 89 million to 382 million barrels annually. The combination of cheap availability and heavy demand offered a straightforward source of state funding. In 1919, Oregon blazed a fiscal trail by instituting a one-cent-per-gallon excise tax on gasoline. The state's motivation was the need for steady financing for the new Pacific Highway and Columbia River Highway projects, which required complex and costly engineering and construction. Oregon collected the tax from wholesalers, who passed it on to retailers, who in turn raised consumer prices at the pump. Other states began to legislate gasoline excise taxes and to earmark the resulting revenue for road work, either through direct accumulation or as security for bonded debt—that is, capital borrowed from private lenders that would be repaid with excise tax receipts.[29]

AASHO took the lead in advocating for the creation of gasoline excise taxes linked to highway construction, and by 1929, all forty-eight states had such taxes. Oil supplies continued to grow—so much so that oil prices fell steadily throughout the late 1920s, allowing states to raise excise taxes despite consumers paying lower prices at the pump. By 1931, the average state excise tax on gasoline was up to four cents per gallon. The result was a regime in which automotive activity and highway development supported one another in a "self-fueling" fiscal cycle. The arrangement offered highway administrators a ready pool of capital that was insulated from political clashes and obstacles: As long as the earmarked excise taxes remained on the books—and they were usually legislated well into the future—they were the purview of the experts in charge of highway development. The success of this linkage between state-level gasoline taxes and road work expenditures did not go unnoticed by the federal government, which added its own one-cent-per-gallon gasoline tax in 1932.[30]

Excise taxes remained out of sight and out of mind, unadvertised and unassociated with rising costs of living, thanks to the low cost of petroleum products. There were no tax revolts and few public complaints. Even oil companies favored gasoline taxes: The revenue produced infrastructure improvements that expanded automotive use that expanded the petroleum market. "Who ever heard, before, of a popular tax?" remarked a Tennessee tax collector in 1926. Perhaps all of this was unsurprising. The taxes were felt minimally by consumers, accumulated rapidly in the state's coffers, and produced popular public goods in the shape of roads and highways. They were also separated from general revenue, a more politically problematic category of taxation. Throughout the 1920s, receipts for state gasoline excise taxes rose collectively from $5.3 million to $431.4 million. And throughout the first three decades of the twentieth century, state road work expenditures rose on average from 2 percent to 25 percent of total state budgets.[31]

For federal administrators, the highway lessons of the early twentieth century were twofold. They had learned how useful excise taxes were, and they had also learned how ineffective state-level highway development was if the goal was to achieve the kind of nation-spanning integration whose absence the 1919 Army convoy had made evident. When economic crisis hit in 1929, the Great Depression began to impart still more lessons. Converging with already changing planning priorities, New Deal relief measures helped set in motion a pivot to urban-oriented, long-range federal highway planning intended to spur the nation's aggregate economic productivity. Indeed, federal administrators understood highways as a politically viable form of relief—but one that fit conveniently within a broader set of nationalizing ambitions. Accordingly, they sought to shift highway policy beyond the limits of farm-to-market planning—still the dominant mode in which highway development operated at the end of the 1920s—and into more comprehensive scales of change.[32]

Following the crash of 1929, even spending-averse President Herbert Hoover explored federal highway aid as an aspect of unemployment relief during his reelection campaign in 1932. Hoover averred in a message to the American Road Builders Association that "road building by the Federal Government and the States and municipalities" was "an important factor, both of relief and recovery." Stimulus spending was a last resort for Hoover, yet deploying public capital toward highway-related employment was a logical fit.[33] If Hoover had to spend public revenue, at least he could funnel it through private business. Despite Hoover's enthusiasm for highway projects, the electorate lost faith in his ability to guide anything resembling "relief and recovery," and Franklin D. Roosevelt won a landslide victory over him in November 1932.

*

Roadworks were well positioned to play a role in the New Deal. While income tax revenue declined by 50 percent between 1927 and 1932, state gasoline excise receipts climbed, hitting a record annual total of $537.4 million in 1931 and rising even higher, to $948 million in 1941, before World War II brought rationing measures and limited the revenue base.[34] The gasoline excise tax revenue continued to provide high yields and was a rare popular form of taxation. The result was a concerted governmental focus, throughout all layers of federalism, on highways. During the New Deal years, relief agencies handled $4 billion in revenue devoted to road work projects.[35]

The first outlines of a comprehensive federal plan for a national highway system came in the Roosevelt administration's vaunted "first hundred days," when the National Industrial Recovery Act appropriated $400 million for highway construction.[36] The goal was to use highway maintenance to ease unemployment. By the next year, however, the Roosevelt administration had begun to conceptualize a broader set of economic impacts that highway development could produce. Roosevelt himself began to see the outlines of an economic growth model built on a modern highway system; new concepts had "really struck his imagination," reported Secretary of the Interior Harold L. Ickes.[37] In December 1934, Roosevelt called a meeting of top cabinet officials and agency leaders—including Treasury Secretary Henry Morgenthau Jr., Federal Emergency Relief Administration Commissioner Harry Hopkins, and BPR Commissioner Thomas H. MacDonald—to discuss the feasibility of a federal highway system that would include "a transcontinental highway from east to west and routes for two or three north and south highways."[38]

Several days later, Morgenthau reported to MacDonald that Roosevelt himself was in the process of drawing up a detailed map of new highways.[39] The president proposed two transcontinental highways. One would run from the Delaware Water Gap in New Jersey to San Francisco, and the other from "half way between Charleston and Savanah [*sic*] running west through Montgomery, Alabama, Baton Rouge and thence to Los Angeles." Three north–south highways would cross these, one running from Danbury, Connecticut, to an unspecified destination in Florida, a second from "somewhere in Illinois roughly to New Orleans" on the east side of the Mississippi River, and a third "from the Canadian border to Mexico running roughly . . . 100 miles inland from the Pacific Ocean."[40]

Like Hoover before him, Roosevelt understood highway construction as a means of unemployment relief. Roadworks offered a useful and politically viable method for easing the jobs crisis, providing a market-oriented public good with federal money that could be channeled through local agencies without raising fears of central state overreach.[41] Boosters, politicians, and road builders already interested in highway improvement welcomed the presence of low-wage, federally subsidized workers. As one governor wrote to Roosevelt in 1933, "There is no form of improvement more necessary and [that] will better serve to relieve the acute unemployment situation than sorely needed highway construction projects on trunk roads."[42] In 1934 and 1936, Congress allocated funds for farm-to-market roads, following the precedents and priorities of the federal highway acts of 1916 and 1921.[43] By 1936, WPA workers had constructed 26,000 miles of such roads, with another 13,000 miles on the way. According to Harry Hopkins, "Farm-to-market roads constitute[d] the greater part of the highway program. . . . WPA has brought the farmer closer to his market and the farmer's child closer to the school. All residents of rural communities have benefited as a result of the farm-to-market road program."[44] The employment provided by roadworks satisfied New Deal relief administrators, who were content that these works created jobs and modernized rural America.

The existing fiscal mechanisms undergirding roadworks provided a welcome flow of capital for relief measures in general. States began to use their gasoline excise tax receipts for non-road-work expenditures, a practice known as "diversion." This was not a new practice, but here its scale shifted. Before the Great Depression, states allocated 2 percent to 3.3 percent of gasoline tax revenue to non-road-work purposes. That percentage surged in 1931, and pushed up to 18 percent by 1937. By that time, concern had mounted throughout the business community. The National Highway Users Conference, representing automotive and roadbuilding interests and chaired by General Motors President Alfred Sloan, successfully lobbied for the Hayden-Cartwright Act of 1934. This legislation aimed to halt diversion, proclaiming it "unfair and unjust to tax motor vehicle transportation unless the proceeds of such taxation are applied to the construction, improvement, or maintenance of highways." Still, the 1934 act contained multiple loopholes that allowed states to continue diverting gasoline excise tax revenue. It required the federal government to link only 60 percent of its gasoline excise tax revenue to roadworks. But from the perspective of federal highway administrators, the act nonetheless ensured a steady stream of politically insulated capital with which to build highways.[45]

The 1934 act gave federal administrators other powers, too, allowing them to spend up to 1.5 percent of federal road work funds on planning, giving them more leeway to design highways as they pleased. In 1936, the BPR coordinated with the states to use this funding to conduct comprehensive, standardized highway use and need surveys—the surveys that led H. S. Fairbank to call for a total revision of highway planning at the end of that year. The surveys offered central state administrators their first clear view of the nation's roads and highways and gave rise to a powerful new planning paradigm. By tabulating statewide highway travel patterns, administrators could now identify whether any given stretch of road could pay for its own construction and maintenance based on traffic—either real or expected—and the revenue generated through gasoline consumption. With anti-diversion legislation on the books, gasoline excise taxes continued to produce new and improved roads, which would cyclically produce more gasoline consumption and tax revenue. The only variable was gasoline consumption itself, but in the 1930s the nation remained awash in cheap petroleum, which promised endless asphalt-borne progress and prosperity. The market had continued to grow, even during the Great Depression: Average annual per-vehicle gasoline use climbed from 525 gallons per year in 1929 to 648 gallons per year in 1941, by which point refineries were earning 62 percent of their income from gasoline sales.[46]

There was a steady supply of petroleum capital with which to build, but highways remained stuck in the relief framework that characterized much of the federal government's activities in the 1930s. Highway administrators began to push different ideas into the public imagination. The highways themselves—and the paths of capital movement they provided—began to hold their attention. More expansive notions of highway-enabled change caught on. Roosevelt's dreams of "smooth, arrow-straight four-lane thoroughfares," noted a writer for the *Washington Post*, might "supply jobs" like traditional road programs. But "super-roads" could also achieve something greater: They could offer "a new form of empire building."[47] More than putting people to work, a federal highway system held the promise of integrating and enhancing the nation's chains of production, distribution, and consumption in a manner that might powerfully transform American political economy.

*

As the 1930s wore on, federal highway administrators made a concerted effort to move beyond a relief framework toward a new model for highway development intended to support the nation's overall economic growth. With excise tax dollars to be spent, officials looked for ways to build a highway system that suited their emerging agenda for linking urban spaces across the country. The process of bureaucratic learning continued: Highways were no longer simply convenient projects that could ease unemployment, but also vectors of change with which the central state could alter the entire nation's social and economic geography—and its government.

When H. S. Fairbank laid out his plans at the 1936 AASHO meeting, he painted the benefits of federalization starkly: No longer would traffic be "delimited by State borders" or "hampered in its proper flow by irrationally variable State regulatory laws" that curtailed effective highway development. Under existing conditions, one state's modern, high-speed highway might turn into another state's dirt two-track at the border. Fairbank called for nothing less than the full "federal assumption of outright responsibility for a Federal system of highways" to solve the problem. He made plain his desire to consolidate federal control. The era's political culture had opened that door wide. "The partitionment of administrative responsibility between the Federal, State, and local governments, long regarded as fixed, has been thrown into question by recent trends," he said, referring to the first term of Roosevelt's presidency. There was a chance to formulate the kind of authority that would be necessary to create a national highway system. He continued: "At one end we have seen the local responsibility in several States yielded in its entirety to the State government, and everywhere questioned as to its sufficiency to meet the modern test. At the other we have seen the Federal government in emergency taking on the local burdens." In brief, there was an opportunity for the BPR to consolidate power.[48]

The states, with ready capital supplies for roadworks, had spent the first three decades of the twentieth century building rapidly and without much coordination. This was the limiting factor of state-based highway construction. It was one thing to be able to build roads; it was another to make them serve broad needs and interests. Federal administrators, and even some state highway officials, began to argue the merits of a nationalized highway system marked by comprehensive planning and smooth route connections.[49] The problem of poorly integrated highways was evident; so too was the challenge this posed to cities and, by extension, national economic exchange. The problem was reminiscent of the early transcontinental railroad system, which developed through several competing private firms, some of which used different gauges of track.[50] While changes in track gauges had once forced trains to stop and passengers to transfer along a route, now the problem lay with easing the flow of commerce from one city to another across state lines. There were problems within cities, too, where limited space and heavy traffic pressed planners to improve urban highways.[51] In time, this issue converged with the racist and destructive practice of "slum clearance," in which highway development became a means to other social ends. At the outset, though, federal officials had not yet even identified the areas in which interstate construction would mean community destruction. Their concerns lay first with the more abstract task of connecting a national web of cities with a highway system.

Throughout the 1930s, cities, like highways, drew federal attention as objects of planning. In 1937 the National Resources Committee (NRC), chaired by Secretary Ickes, devoted a lengthy report to the topic. "Our Cities: Their Role in the National Economy" made the case for the kind of highway-based connectivity elaborated by federal highway administrators. Nearly two decades had passed since the 1920 census found that, for the first time, the majority of Americans lived in urban areas. Now, the NRC argued that the nation's collective urban space served as the "workshop of our industrial society and the nerve center of our vast and delicate commercial mechanism." The nation's ninety-three cities with populations over one hundred thousand now accounted for over 75 percent of all wholesale exchange. Yet despite the centrality of urban areas to American life, problems abounded. Chief among these was that "competing forms of transportation [had] left their disrupting imprint upon the national urban pattern." In terms of industrial production, railroads had become expensive in comparison to truck transportation. But, more fundamentally, the report blamed the federal government for leaving the highways on which urban interdependence relied "uncoordinated."[52]

Highway officials were aware of this lack of coordination. The NRC was finalizing "Our Cities" as the pivotal 1936 AASHO policy meeting convened. While early plans discussed with President Roosevelt had set the stage for changes to come, it was not until this event that a recognizable description of the highway system we travel today began to crystallize. Within a decade, the vision articulated at the meeting gained legal force with the Federal-Aid Highway Act of 1944 and its outline of a 40,000-mile "National System of Interstate Highways" capable of linking the nation's urban economic centers.[53]

Soon after Fairbank spoke in 1936, BPR Commissioner MacDonald took the stage in San Francisco. In a speech entitled "Looking Toward the Highway Future," MacDonald drew on his knowledge of European highways to elaborate the vision Fairbank had articulated. There were many lessons to be considered. For example, in London, MacDonald had previously observed the construction of the "Great West Road" (now the A4). Its legislative origins lay in providing employment, but it offered a "wide thoroughfare" for traffic traveling between the city and surrounding "suburban and rural districts." When MacDonald first saw the roadway in 1928 it was "bordered by open fields." Now, he reported, it was "lined by continuous large industrial enterprises." Marred by congestion, it had "lost its visioned function to move traffic expeditiously." The lesson MacDonald conveyed was that "population and industries" gravitated "rapidly and certainly" to highways.[54] The Great West Road, in other words, was designed to accommodate traffic needs extant at the time of its construction—not to accommodate the growth that, evidently, followed its arrival. Highways needed to be built for the future rather than for the present.[55]

MacDonald found a better model for future-oriented planning, and the administrative governance necessary to achieve it, across the English Channel. France's National Department of Highways trained its officials at the famed École nationale des ponts et chaussées (National school of bridges and highways). While the Great West Road in London had offered a case study in how highways altered social and economic arrangements, France provided broader insights. French officials had advanced "two distinctive modes of highway planning": first, "an intercepting circular highway enclosing the city, and second, motor highways, radial from the city, reaching for a considerable distance beyond to connect with existing national highways." Beltways around urban areas and the spoke-like highways that pierced them offered "permanent relief" by dispersing businesses and housing. Officials could add concentric rings of interconnection as urban and suburban space expanded. MacDonald determined that the French system would obviate the problems he had found along the Great West Road.[56]

While France offered MacDonald a spatial vision, he was still in search of a governmental method for making that vision a reality. It was in Germany that MacDonald's plans for the US matured. MacDonald claimed that Germany, much like the US, had long built roads on a "local basis." This changed when, he explained, the "National Socialist Government took over the development, in 1933, of a national system of highways." MacDonald drew his information from Robert Allmers, president of the German Reichsverband der Automobilindustrie (Imperial Federation of the Automotive Industry). Quoting Allmers at length, MacDonald drew a parallel between the German and American situations. The German landscape had been marked by "a hopeless state of disintegration" akin to the "Middle Ages." According to MacDonald's paraphrase of Allmers, "State and provincial governments, district and communal authorities, made every effort to obstruct a sound development based on uniform principles. . . . Every district road engineer built his roads in a different way, but nearly all of them built them in the wrong way."[57]

Germany implemented a two-part solution. Overseen by the inspector general of German roads, states and provinces repaired and improved some existing highways. More significantly, the inspector general's office planned and built "a wholly new system of roads known as the Reichsautobahn," consisting of three north–south and three east–west routes across the country. "The literal translation of the word Reichsautobahn," MacDonald reminded his audience, was "national auto road"—though this was perhaps a more strategic than literal translation that MacDonald invoked. Officials in charge of the Reichsautobahn focused on planning "ahead of the highway traffic," in a manner capable of accommodating future transportation. Even more than

France's radial approach, this was the "very antithesis" of the existing American approach to highway planning. MacDonald's experiences in France and Germany led him to believe that "over a long period, a highway system originally well planned can be expanded to meet the growing needs [of the population] without large abandonment of investment or changing to wholly new policies." If London had offered a cautionary tale, France and Germany seemed proof that modernist rationality could account in advance for the contingencies of growth.[58]

But how to bring this about in America? MacDonald puzzled over the economics of Germany's highway development approach. It was in part, he thought, intended to boost employment, much like traditional roadworks in the US. Yet German highways seemed to play a broader role in growth. MacDonald came to the conclusion that there was "no immediate possibility" of Germany recovering the costs of highway construction "directly from imposts on road users." Germany had not followed a toll-based approach; the Reichsautobahn would not simply pay for itself through fees for its use. Instead, MacDonald surmised, it would pay for itself through its general contribution to the German economy. And Germany seemed to harbor no qualms about this approach. MacDonald noted Germany's efforts to incentivize automobility, making "every inducement" "to encourage the development of motor traffic." Unlike the rest of Europe, Germany stood out for reducing fees and taxes associated with automobiles. The result was an enormous leap in licensing, from 41,000 cars in 1932 to 180,000 in 1935. The automobile industry had also been nationalized in this period. All in all, MacDonald found "an unusual expression of faith in the utility of highway transport"—and a model for national administrative and economic growth brought about with the help of a national highway system.[59]

MacDonald doubled down on the lessons he learned in Germany. "We must grasp this highway problem in this country more firmly," he concluded, turning his attention back to the American context. "We must raise our standards to the new levels demanded by the universal utility of the motor vehicle."[60] A futuristic highway system awaited: "Its surface will be consistently smooth, and nonskid." It would have "shoulders wide enough for stopping; no steep side ditches." And in places with heavy traffic, there would be "four-lane roads" consisting of "two lanes on each side of a center parkway."[61] It would link America's metropoles in an efficient, rational network of productivity and prosperity that would impress even the Germans. To achieve this, MacDonald recommended, "two distinct programs" should be carried out by administrative agencies: first, a "systematic rehabilitation of existing highways" sufficient to make them capable of sustaining heavier traffic volumes well into the future, and second, new "belt line intercepting highway[s] plus radial roads on new right of ways to serve the metropolitan areas." This highway system would go "a step further" than even the German highways and "provide for the complete separation of local from through travel."[62] If Fairbank had offered the conceptual model for a network of cities and economic growth, MacDonald began to draw the physical and administrative blueprint necessary to make that vision reality.

*

MacDonald's 1936 pitch for centralized long-term planning capable of meeting future traffic demands came alongside the BPR's attempts to enhance its expertise and capacity. Under MacDonald's leadership, the BPR began to invest heavily in its research division. Sophisticated soil science was the key to building durable roadbeds capable of supporting a uniform national highway system throughout the continent's diverse ecological contexts. Understanding that "the character of soil on which a road surface is placed is often as important as the type of surface itself," researchers began to study the various capabilities of "bituminous material, Portland cement, slag, limo, limestone dust, common salt, calcium chloride and soap" in various climates.[63] The goal was to engineer substrates for all weather conditions. Officials even adapted seismograph technology used in oil prospecting to detect solid rock beneath proposed roadbeds.[64] Above ground, BPR officials used new camera technologies to understand driver behavior.[65]

While scientific experimentation spurred technological advances, the BPR expanded its reach through subnational administrative units that surveyed national automotive trends and produced a composite portrait of the benefits of a national highway system. Comprehensive highway planning surveys began in 1936. Making use of state highway departments and their personnel, the BPR collected "information to put future highway improvement on a sound, business-like basis." The surveying work captured highway conditions, traffic patterns, and financing with the intention of enabling national highway development to proceed "based on economic and social needs."[66] This was the first step in drawing up new construction blueprints. With a nation's worth of aggregate highway data in hand by 1937, the BPR then assembled a team of highway engineers to institute uniform highway design standards. The engineers were AASHO members and former employees of state highway departments, now tasked with bringing highways within the purview of the federal government.[67] To house this massive push into federal highway research and development, the BPR opened a $1 million research center covering fifty-four acres in Abingdon, VA.[68]

The BPR flexed its growing administrative strength in Congress. In 1938, the Senate blocked a massive $8 billion aid bill for the construction of toll-funded highways—a bill that originated with relief-focused politicians rather than with the increasingly professionalized administrators of the BPR. The bill's supporters framed and justified it as a jobs program and claimed the resulting highways would be self-sustaining based on subsequent toll revenue. But as BPR administrators argued before the Senate—drawing in part on their 1936 highway surveys—there was no evidence to support the financial viability of a tolled system when nontolled alternate routes existed. Nor had the bill resulted from a study of transportation needs; instead it focused on unemployment relief.[69] By channeling funding through relief agencies rather than highway departments, the bill would disrupt established road work partnerships between the federal government and individual state governments. Emphasizing the professional expertise of highway administrators, MacDonald maintained that "one of the worst features of the bill" was that it "by-passe[d] all existing, experienced agencies such as . . . State [highway] departments," as well as the BPR itself.[70] Apart from providing temporary employment, the project threatened to be a financial and bureaucratic disaster. And, in a move intended to foreground its own emerging highway plans, the BPR noted that the 1938 bill did little to rectify the "unimproved" nature of highway connections between the nation's cities.[71]

*

When the 1938 highway bill died in the Senate, the BPR succeeded in replacing the unemployment relief framework with a new focus on national development. Congress soon requested an official study from the BPR on the feasibility of a more carefully planned toll-based national highway system. *Toll Roads and Free Roads*, the BPR's resulting report, appeared in 1939 and reached several significant conclusions. First, it fully rejected the possibility of a toll-based highway system, supporting the position BPR officials had taken in 1938. Second, the report offered an alternative "master highway plan for the nation," proposing a system of "free" highways funded with excise tax revenue. Third, it centered national security concerns, arguing that free highways would meet "the requirements of the national defense and the needs of a growing peacetime traffic," identifying both as pressing issues in "metropolitan areas." The report consolidated key trends in federal highway planning by arguing against toll financing, making claims about national defense, and aiming to connect urban spaces.[72]

Avoiding toll financing was the preeminent focus of *Toll Roads and Free Roads*, as its name implied. To make the case against toll financing, the report's authors drew up two contrasting plans for highway development. The first was for a 14,336-mile-long toll-based system that might collect revenue *almost* sufficient to recoup construction costs. This plan followed the instructions of the Federal-Aid Highway Act of 1938, which had directed the BPR to "investigate" routes for six "superhighways"—three east–west, three north–south—funded with tolls. Even after selecting the most toll-viable pathways, the BPR could identify only one stretch of this system, a 172-mile segment connecting Philadelphia and New Haven, that could be expected to carry enough traffic to pay for itself via toll charges. An array of statistical evidence drawn from the 1936 state highway surveys clarified the problem with toll financing as well as with a national highway system limited to six superhighways: Most travelers would not pay tolls when parallel free roads existed, even if ease and speed of travel were at stake. The BPR's extensive state-by-state surveying indicated as much, as did a poll conducted by Gallup, which found in 1938 that 81 percent of Americans answered "no" when asked if they would pay tolls of half a cent per mile to fund national highway development.[73]

These data conflicted with contrary examples, albeit at smaller spatial scales. The success of the New York State Thruway and the Pennsylvania Turnpike suggested that tolls could fund advanced highways.[74] But the dynamics of those two toll-based state highways offered inverse support for federal highway administrators' ideas about toll-free development. The Thruway and the Turnpike navigated wealthy, industrial, and densely populated regions.[75] The model they offered was not exportable to, say, South Dakota's sparse expanses. The same went for several other low-density regions. And whatever their successes, the Thruway and the Turnpike did not achieve national integration. The Thruway, for instance, connected Buffalo and New York City, key industrial cities at opposite ends of an otherwise rural state. But shifting to a broader scale and connecting farther-flung places such as New York City and San Francisco—a common dyad in highway administrators' visions—required a highway system that traversed landscapes where traffic volume and toll revenues would be insufficient to enable toll-based construction.

Turning away from the congressionally mandated proposal for six toll-based superhighways, federal highway administrators sought to release highway development from the spatial requirements of toll financing. Many goals had guided federal highway administrators leading up to the *Toll Roads and Free Roads* report, from concerns raised by the 1919 Army convoy about the nation's vast and unnavigable interior to Fairbank's focus on building "direct connections" between distant cities to MacDonald's call to transcend the "local basis" on which existing highways had taken shape. All of these goals shared in the notion that the nation and its economic progress required new patterns of automotive movement. And the BPR recognized that such patterns could not be inscribed without an enabling financing mechanism, one that freed highway administrators from existing "impediments" to planning and construction.[76]

The second half of *Toll Roads and Free Roads* presented "A Master Plan for Free Highway Development," which offered an array of evidence making the case for a highway system financed with taxes and bonds rather than tolls. Tolls required existing or at least highly predictable future traffic to fund highway construction and maintenance. Other revenue collection strategies, in contrast, might allow administrators to separate capital from existing infrastructure, offering them the fiscal flexibility to place automotive paths wherever they wished.

A point of fracture had revealed itself: The spatial and economic aims of a national highway system did not match existing arrangements—not in fiscal, physical, or administrative terms. With a concerted emphasis on avoiding tolls and linking cities, BPR administrators called for a federal system of "interregional highways," planned in accordance with "the States and the War Department." The report proposed a 26,600-mile-long system of these "free" highways—underscoring the absence of tolls—capable of providing "connections through and around cities." This highway system would even "meet the requirements of the national defense"—a concern highway administrators increasingly used as a political ploy to boost their goals. Escaping the constraints of toll-friendly geographies would extend highway mileage and herald "modernization" by connecting the nation's large population centers.[77]

Recalling MacDonald's study of France and Germany, the report called for "belt-line distribution roads around the larger cities and bypasses around many of the smaller cities and towns." Motorists would funnel into metropolitan areas via "arteries" that averted "the countless impediments that embarrass the movement of twentieth-century traffic." Long-range traffic between metropolitan areas would move swiftly and directly, unhindered by towns and villages. Within urban centers, BPR officials claimed their plan for "unfettered circulation of traffic" would reverse the "slow decay" that had become characteristic of the average "business district." What would become of communities cut off from the interregional system, they did not say: Their emphasis rested on the "great convenience" motorists would experience if, instead of passing through such communities, they moved directly from city to city.[78] With the 1939 report, administrators explored their economic and geographic visions freely, setting aside the constraints of existing infrastructural systems and social arrangements. "A Master Plan for Free Highway Development" offered an alternative to the insufficiencies of toll financing, outlining a method for simultaneously linking disparate cities and conquering the vast interior expanses that had concerned military and economic planners in 1919.

Yet despite the clarity offered by the 1939 report, it only stalled discussions of highway planning and financing on Capitol Hill. While MacDonald and the report's other authors saw no viability in financing a national highway system through toll charges, many politicians maintained that highway projects paid for in such a manner would gain more public support than highway projects paid for by less direct tax mechanisms—even equally user-based taxes collected in the form of excises on gasoline, oil, or other automobile-related commodities.[79]

*

While a national highway system remained financially stalled in the late 1930s and early 1940s, officials continued to push for federal highways where they could, now against the backdrop of war. Highway planning evolved alongside other federal efforts to manage the wartime economy. In December 1940, Secretary of War Henry L. Stimson convened the Highway Transportation Advisory Committee (HTAC). Chaired by MacDonald, the committee included representatives from the American Association of Motor Vehicle Administrators, the American Association of State Highway Officials, and the International Association of Chiefs of Police. Together, their task was to inform the War Department of optimal domestic military transportation routes, to collect the records of civilian drivers, and to inventory the nation's trucks and buses.[80] Like the military officials involved in the 1919 Army convoy, HTAC leaders recognized the convergence of military and civilian highway possibilities. As early as the spring of 1941, the Roosevelt administration began to frame federal highway planning as a tool of eventual reconversion to a peacetime economy. President Roosevelt requested highway plans to be "put on the shelf, ready to take out when the end [of the war] comes."[81]

Highway officials realized that war would be a boon for their ordinary business and a useful context in which to advance their long-standing goals. They adapted to war-focused political culture. AASHO held its 1941 annual policy conference two months before the bombing of Pearl Harbor, and its members had already begun to articulate policy in terms of contributing to the war effort. G. Donald Kennedy, Michigan's state highway commissioner, told the assembled officials that "the fullest utilization of highway transport" would be "indispensable" to America's "Arsenal for Democracy."[82]

Kennedy, like so many other administrators who led the era's conceptual revolution in federal highway development, began by painting a grim picture of existing highway infrastructure. The problem, he said, was that federal highway support had arrived "rather late." Aid remained sparse relative to the cost of modern construction, and the war was making matters worse. As recently as a month previously, Kennedy's own Michigan State Highway Department had considered laying pavement without steel reinforcements and had even evaluated plans for wooden bridges. The concern was that the state would not have sufficient capital to buy steel, and other materials presented problems, too. Tar, rubber, and paint—all critical paving and maintenance components—were in short and expensive supply as the war ramped up and diverted these critical goods from domestic uses. Increasingly, Kennedy reported, state officials had no recourse but to turn to the federal government for aid in attaining materials.[83]

For state officials like Kennedy, much could be achieved by federalizing highway development. What states might lose in formal authority they would more than gain in fiscal benefits, and it was unclear whether they would lose any authority. Wartime made for a compelling new context in which to articulate new ideas about federalism and related financial benefits—and lobby for a national highway program. Concerns about defense and security, of course, held the nation's attention. But Kennedy argued that Americans "tended to over-emphasize" differences "between emergency and continuing highway responsibilities." As he put it, "much of the work undertaken to meet the emergency can be made to serve the permanent needs of civilian traffic." There were real military highway needs to be met, and meeting those needs would also assist the national economic framework.

The crisis of war converged with a "growing understanding of . . . the important social and economic benefits that can be effected by wise highway development."[84] Kennedy asked his audience to consider the case of Ypsilanti, Michigan, where an Army bomber plant planned to transition sixty thousand men from automobile manufacturing to war production. But the plant was twenty-five miles from Detroit and Dearborn, "the only considerable reservoirs of labor in the area." Two options presented themselves: A "migrant industrial population" could be "housed in a mushroom settlement in the immediate neighborhood," or new highways could be built to make the plant "accessible" to Detroit and Dearborn residents.[85] The state chose the latter option, and the Michigan State Highway Department worked with the Public Roads Administration (PRA) to design local arterial roads for effective commuting. (The PRA was the same entity as the BPR. The agency had been renamed in 1939 but would revert to the BPR title in 1949 and eventually, in 1966, be transformed into the Federal Highway Administration.) Kennedy predicted that the system would eventually "fulfill the urgent peacetime needs of the Michigan metropolis"—a sure bet, given that the Ford Motor Company operated a 63,000-square-foot plant in Ypsilanti. More broadly, the highways would constitute a "vital transport route between Detroit and Chicago."[86] He believed his case study applied nationally. The bomber plant was just one example of defense needs producing "permanent contributions" to "normal highway needs" and "regular highway programs."[87]

The relationship between military and civilian highway planning grew stronger. In November 1941, Roosevelt signed the Defense Highway Act and created the Office of Defense Transportation.[88] Following the work of the HTAC, the bill enhanced funding and administrative capacity. Much of this support went to highway surveying, with states receiving $25 million to study and repair "critical deficiencies" in the nation's "strategic network of highways and bridges." With a federal-to-state funding ratio of 75–25, the 1941 act made an unprecedented federal financial commitment to highway development and shifted additional planning power to federal officials.[89] Explaining the changes, Thomas MacDonald noted that the funds were apportioned to all forty-eight states, the District of Columbia, and even the territories of Hawaii and Puerto Rico.[90] No construction would begin until the end of the war, but final blueprints would be ready and waiting, with a particular focus on plans for a "network of highways and bypasses around and extensions into and through municipalities and metropolitan areas."[91] The growing conceptual emphasis on urban highways was now reflected in legislation—as was a precedent for expanding federal funding through national security claims.

The Roosevelt administration also established the National Interregional Highway Committee in 1941. Its name referenced the key proposal in *Toll Roads and Free Roads*, and the new committee's members examined traffic data, route planning, and financing mechanisms necessary to advance a national highway system. The committee's specified role was to advise the Federal Works Agency, headed by John M. Carmody.[92] Among the highway experts from a range of agencies who filled the committee were Kennedy and MacDonald. As Roosevelt reasoned to MacDonald, such preparation "will permit us, upon the conclusion of our defense program, to utilize productively some of the man power and industrial capacity then available to construct a national system of interregional highways."[93] Roosevelt's statement reflected fears of a return to economic depression when the war finally came to an end. But the emphasis had shifted decisively. Unemployment relief had once been an end in and of itself when it came to roadworks. Now the construction of a national highway system held its own position in federal policymaking—a position that might help manage employment, certainly, but a position no longer defined by local issues or job creation.[94]

The emergence of federal highway planning as a distinct realm of policymaking came with sustained rhetorical support and explanation from administrators. "In this dark night of highway transportation," as H. S. Fairbank (now deputy chief of the PRA) characterized the war, "one interested in the well being of our highway system and the future of highway transportation . . . turns his eyes toward 'the dawn of peace'" and "asks . . . for a timely preparation to deal with the highway problem sure to come." Delivering federal updates to a group of road builders in 1942, Fairbank spoke of the "epochal events" sure to follow the war's end, and of the "magnitude and difficulty of the many changes that peace will bring." He discussed planning trends that would soon be codified in the Interregional Highway Committee's report. Though his rhetoric shirked his own agency's responsibility in the matter, he specified that it was in and around metropolitan areas that "we shall need in the post-war period to build our superhighways."[95] As new conceptions of the economy took root alongside the exigencies of war, the PRA's approach to federal highway development took hold, too.

In another decade's time, Fairbank's ideas became reality. And his reflections on the changes afoot prefigured the findings of later urban historians. As he put it, "The highway in the past has done some disservice to the city. In many ways it has helped, to be sure; but in its clumsiness it has injured, too. It bears a large share of responsibility for the city's deca[yed] areas. It has lured taxpaying homeowners to homes outside the city boundaries and left the city with the burden of dealing with the areas those taxpayers vacated."[96] At the time, Fairbank predicted optimistically that a carefully planned federal highway system would avoid causing such problems in urban areas in the future—a prediction that failed spectacularly. One thing was clear, though: When peace came, highways would occupy a new "place in the mind and habits of the people," as well as "in the new economy of the world." This, he said, was a "foregone conclusion."[97]

*

The National Interregional Highway Committee published its study, *Interregional Highways*, in January 1944 and launched a decisive shift to urban-centered federal highway planning. The report followed a now familiar narrative: Naive early road builders, with myopic local visions, had filled the nation with muddy farm-to-market byways. Now, the nation needed a "vast road network" that gave "preferential order to the improvement of the portions of greatest use potential"—a stodgy, politically calculated description of urban highways. The report expanded the arguments made five years earlier in *Toll*

Roads and Free Roads, and PRA administrators offered a map for a national highway system predicated on boosting the productive capacity of metropolitan areas and linking them to one another for broad economic benefit. The routes looked similar to those in the 1939 proposal but added still more mileage in and around large cities.[98]

The Federal-Aid Highway Act of 1944, signed into law by Roosevelt on December 20 of that year, followed the recommendations made in *Interregional Highways*. With the end of World War II expected, Congress had turned its attention to the challenge of shifting the nation back to a peacetime economy. In this regard, infrastructure development looked ready to play a role in structuring investment and production models capable of taking the place of war production.[99] Roosevelt framed the 1944 act as a necessity to prepare both for the return of soldiers who would need work and for the resulting era of peacetime affluence for which all hoped. Highways, Roosevelt made clear, demanded "prompt and vigorous attention." They were, he emphasized, an "essential . . . part of an expanding, prosperous economy" to come.[100]

The 1944 act dispersed a record $500 million in federal roadworks funding over the next three years. Directed toward existing highways as well as new highway projects, the enormous appropriation further established highways as objects of federal interest and activity. Congress apportioned the funds following an "ABC" classification system. "A" highways comprised the "primary system" of highways designated for federal aid in previous legislation dating back to 1921. This would receive 45 percent, or $225 million, of funding under the 1944 act. Another 30 percent, or $150 million, would go to the "B" or "secondary system," mostly farm-to-market roads and other routes in rural areas and within municipalities with populations under 5,000. And—marking a radical shift—25 percent, or $125 million, would support the "C" system of newly demarcated highways and arterial connections around cities. "C" highways received the $125 million in addition to any funding already earmarked for urban roads within the "A" system. Reverting to the 50–50 prewar federal-to-state matching funds ratio, individual states would contribute equal financing to each of the three categories.[101]

The 1944 act also marked the most profound change yet in moving administrative authority from the state to the federal level. The apportionment process specified in the new statute ensured that each state's matching funds would go where federal officials wanted them, and as the funding schedule made clear, this meant urban areas.[102] The act also allocated federal funds for rights-of-way acquisition, which had previously been left to state-level

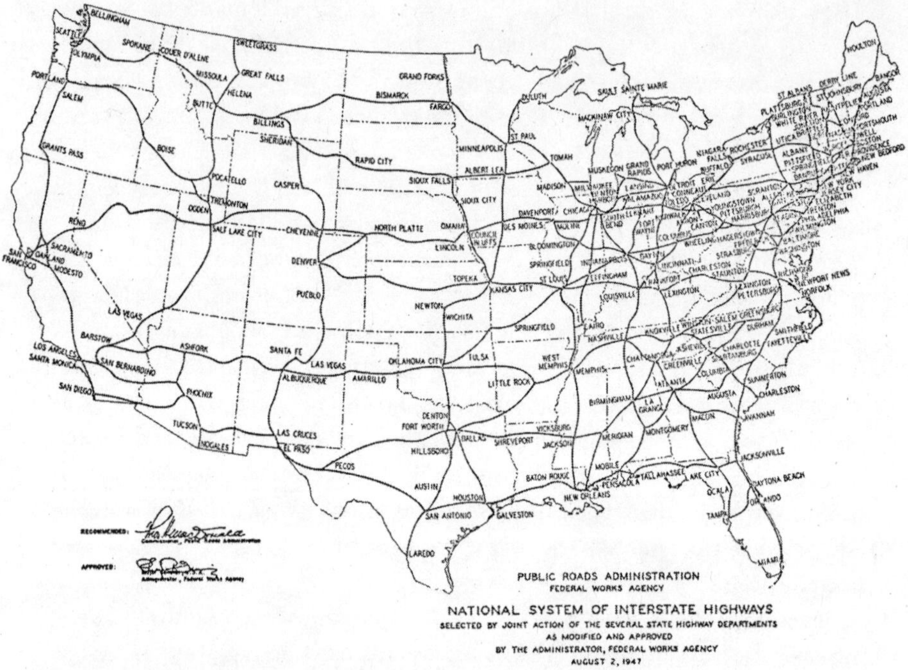

FIGURE 4. The National System of Interstate Highways as designated by the Public Roads Administration, 1947. Courtesy of the National Archives and Records Administration.

financing. This facilitated projects in urban areas, where rights-of-way tended to be far more expensive than in rural areas, and it ensured that federal administrators would oversee urban route selection. The 1944 act also located regulatory power for "any highway or street hereafter constructed with Federal aid in any state" at the federal level, and specified that the "location, form, and character" of the most basic details—from traffic signals to pavement markings—would require the approval of the overseeing federal agency as well as the relevant state highway department. The 1944 act gave the federal government a role in all aspects of highway development, from planning to financing to construction standards.[103]

And most significantly, the 1944 act mandated the creation of a 40,000-mile "National System of Interstate Highways." Anticlimactically, Congress delivered no funding for the nascent interstate system, limiting financial outlay to the existing A-B-C highways for the time being.[104] More than a decade would pass before funding arrived for the interstate system and its construction began. In the meantime, federal officials set themselves the task of selecting routes for their pathbreaking, if still only cartographic, federal highway system. The PRA announced its final plan for interstate highway routes on

August 3, 1947, delivering a map that connected 135 cities with populations above fifty thousand, as well as thirty large metropolitan areas. In and immediately around these urban contexts, 2,300 miles of interstate routes were designated for further study. But the majority of the interstate system, amounting to 37,700 miles, was ready to become a reality.[105]

Construction would begin as soon as the federal government appropriated capital and deployed bulldozers. Bureaucratic learning would continue, but the conceptual revolution that had transformed highways into technologies of central government was complete. Soon enough, the highway-borne governing regime would begin to distribute its benefits and burdens in durable, uneven patterns. As officials cleared paths for the massive new infrastructure, highway construction and community destruction were inseparable processes. The new highways offered mobility to some and presented obstacles to others when they demolished urban neighborhoods and sliced up farmland. But before those problems commenced, the federal government had to find a way to finance this vast new infrastructure project. In 1936, Fairbank had warned that "the cost of a highway system is a perpetual one; that the limit of new highway building is reached not when we have built as many miles as we might be willing to pay to build, but when we have built as many as we are willing to pay indefinitely to keep." In the years ahead, men like Fairbank worked hard to tie the interstate system's financing to taxes on ever-mounting petroleum consumption. This, they hoped, would ensure that capital would not limit their dreams. But appropriating billions of dollars—and finding ways to keep the appropriations rolling well into the future—would require considerable congressional wrangling. And even when highway officials got the bottomless pit of money they wanted, it was not to be free.

2

Primed at the Pump

The afternoon of June 29, 1956, found President Dwight D. Eisenhower counting down the final twenty-four hours of a three-week stay in Walter Reed Army Medical Center. He had been hospitalized on June 7 with a severe bout of ileitis that required surgery. The intestinal ailment—rumored to have resulted from eating too many "greasy tidbits" at a party—worried a nation that knew the president had survived a heart attack just the year before. So, in a show of strength, Eisenhower spent his final day in Walter Reed meeting with Richard Nixon to plan the vice president's upcoming mission to the Philippines and signing twenty-seven bills that had accumulated during his hospital stay. Chief among these was the Federal-Aid Highway Act of 1956, set to fund the long-awaited interstate highway system and transform its maps and blueprints into the most expensive and expansive public works project in American history.[1]

Popular accounts of the interstate highway system's origins tend to credit Eisenhower personally with its development. As the story goes, Eisenhower shepherded the enormous public works project into existence as a beacon example of beneficent, bipartisan state building that signaled national strength and preparedness at the peak of Cold War fears of nuclear attack.[2] Of course, the interstate highway system had existed in law for nearly a decade by the time Eisenhower took office—the Federal-Aid Highway Act of 1944 had outlined a "National System of Interstate Highways" to connect "principal metropolitan areas" and "serve the national defense." The 1956 legislation moved "defense" into the act's title and added 1,000 more miles to the system, making a total distance of 41,000 miles. Still more significantly, it added fiscal provisions capable of turning mandates into physical infrastructure and made the federal government responsible for 90 percent of construction costs. Convalescing in his hospital bed, Eisenhower declined to be photographed as he signed the bill. The lack of fanfare helped to veil what should have been obvious: By many measures, the plan he favored for the interstate highway project had failed.[3]

Rather than an artifact of Eisenhower's statesmanship, the Federal-Aid Highway Act of 1956 was a product of a sharp partisan battle in Congress over the project's fiscal and physical contours. It is true that most politicians in the 1950s supported the idea of a national highway network, and the 1956 highway act arrived with a mix of Republican and Democratic support. But that support resulted from power-wielding as much as deal-making. Following the 1954 midterm elections, Democrats gained majorities in the House and Senate. They then held a line against Republican financing plans and backed Eisenhower into the political corner in which he signed the 1956 highway act, with their own fiscal plans at its center.[4]

*

As momentum built around the interstate highway project in the 1950s, legislators had to answer a pivotal question: Who should pay for the interstate system, and how? By the early twentieth century, local and state bond issues had come to constitute a common source of funding for public works projects. This trend accelerated around the midpoint of the twentieth century. In the post–World War II years, low interest rates led local and state officials to issue bonds readily, considering debt to be a cheap source of capital. Historians have devoted significant attention to subnational bonded debt in this period. But the role of bonds at the federal level in these years has received less scrutiny. In the case of the interstates, bonded debt provoked enthusiasm and concern in equal measures, producing a rare mid-century instance in which objections to bonded debt not only gained political articulation but prevailed.[5]

Concerns about bonds were not unwarranted. A government-issued bond can be thought of as a profit-seeking private investment in public development. A bank or other institutional lender delivers capital to a unit of government—whether local, state, or federal—for use on a specific project. The governmental borrower owes the lender the principal sum borrowed, as well as interest at an agreed rate, payable over a specific time frame. The bonded debt is secured by tax dollars: The principal and interest to be repaid derive either from revenue collected from the use of the project funded by the bonded debt, or from general tax revenue.

Despite carrying the cost of interest, bonds offer governmental borrowers two key advantages. First, borrowing capital delivers cash up front, avoiding the need to slowly accumulate tax money before there is a pool of capital sufficient to begin a project. Second, bonded debt avoids the need to raise taxes or levy new ones to accumulate revenue. Amortization—the process of paying off principal and interest across time—in the long run costs taxpayers money beyond the amount expended on the project in question. But tax rates can remain stable throughout amortization, because tax revenue can continue funneling to lenders even after a project is complete. Private capital delivered as bonded debt, in other words, can enable low tax rates even as tax revenue piles up in excess of the amount of capital needed for a particular project. Additionally, bonds support the long American tradition of keeping government "out of sight," by blurring distinctions between "public" and "private" actions.[6]

In the 1950s, these aspects of bonded debt appealed to Republican officials, who hoped to use this financing method for the interstate highway project. Debate over using bonds in this way came amid broader political arguments about government spending. In the years after World War II, according to historian James Sparrow, it became clear that Americans "had authorized a government far larger and more intrusive than the New Deal state had ever been." For some, this was cause for alarm, and the 1952 presidential election became a referendum on the situation. The United States was at a "crossroads," charged Senator Robert Taft, the conservative firebrand known as "Mr. Republican," as he campaigned for his party's nomination. Thanks to the "economic plans of the long-haired braintrusters"—a reference to New Deal interventions—"our liberty is threatened by big government." While Taft's presidential aspirations were cut short by Eisenhower's primary election victory, Taft continued to call on "every American" to support Eisenhower in the general election and restore a government built on "honesty, efficiency, economy, and above all liberty." Eisenhower spent his campaign decrying "excessive taxation" and the "ceaseless expansion" of the federal government, a source of "evil" that would "throttle" freedom.[7]

Paradoxes soon emerged. Conservatives were not, it turned out, averse to using government to achieve their desired goals. Once president, Eisenhower expressed enthusiasm for directing federal power toward reordering "the very structure" of American life with the interstate highway system.[8] But it was hard to square that desire with the Republican Party's tax-averse positioning. And it was difficult for conservatives to explain their support for a massive public works project that facilitated active governmental management of society and the economy. Administration officials became determined to pay for interstate construction with private capital borrowed from banks. Bonds offered a logical way for conservatives to support the interstate highway project while maintaining their antipathy toward even the most incremental of tax levies and governmental activities. With bonds, they hoped they could frame private profit-seeking as the true engine of national progress.[9]

Bonds could deliver capital quickly and solve some of the political challenges of conservative state building. But, as Democrats argued, bonds indebted citizens and their representatives to profit-seeking institutions, raising questions about the political economy such an arrangement produces. Bonds required payments from present and future taxpayers and obligated public revenue to the private sector. After all, bonds did not erase the need for taxation, but merely diverted tax dollars from the public purse into private coffers, where they accumulated as profit. Democrats consequently worked in Congress to replace bonded debt with direct revenue expenditure drawn from excise taxes on gasoline and other automotive goods. Excise taxes—levied on the sale of a particular commodity—offered a straightforward relationship between taxpayers and government activity.[10]

Broad political support for the interstate highway project belied just how forceful the debate over its financing would become. At stake were competing social and financial visions of how government should provide public works. Since 1936, federal highway planning had been premised on spurring the "national economy."[11] It was easy to assume that aggregate growth and widespread prosperity in people's day-to-day economic lives were the same thing. Taken with such ideas, elected representatives and the public at large agreed with the basic bureaucratic plan, which became the first legal instantiation of the interstate highway system in the Federal-Aid Highway Act of 1944.[12] The administrative consensus that federal highway development, though costly, would boost the nation's overall economic productivity had lodged in the halls of power. In 1954, Eisenhower proclaimed that the interstate highway system would be the key means of enhancing "interstate commerce" and national productivity.[13]

But the financing problem remained, made all the more difficult by the magnitude of the endeavor. It was not just a matter of deciding how to raise capital. The scale of the interstate highway project and the conservative politics of the 1950s meant that the problem of national debt had to be reckoned with, too, before moving interstate highway legislation forward. Just as ideas about economic growth had shaped the cartographic design of the interstate highway system two decades earlier, national accounting concerns now influenced its financial design. It was not just GNP that prodded at lawmakers but concerns about aggregate expenditure. The Eisenhower administration, espousing a deficit-averse doctrine of robust economic growth *and* balanced budgets, began to search for a financing approach consonant with its claims about how government should spend money—or rather, *not* spend money. This was a difficult task: The $27 billion projection for interstate highway construction was an enormous line item—over $1.6 trillion in today's money. How could Republicans advance such an enormous and costly project while maintaining their defining image of fiscal responsibility?[14]

*

The political and rhetorical challenges came on display when Vice President Richard Nixon delivered the Eisenhower administration's first pitch for the interstate highway project to the 1954 National Governors Association conference in upstate New York. After opening with pleasantries and jokes, Nixon posed a simple question to the crowd: "Where is the United States going, and by what road?" The metaphor was apt.[15] According to Nixon, a forked path loomed. Turn one way, and the nation would lose itself in a thicket of taxes and government projects leading along the "back road to socialism," as the US Chamber of Commerce put it. Turn the other way, and the nation could embrace socialism's antipode, "free enterprise." The choice, said Nixon, was clear. And the way to realize that choice was via "a grand plan for a properly articulated [highway] system," an ambition that he said must receive "top priority." Nixon pulled no punches in explaining at length that the state of the nation's "highway net" was "inadequate locally, and obsolete as a national system." Existing highways were an anachronism; they had "just happened"; they were "haphazard," "completely arbitrary," and suitable only for "local movement at low speeds." They would not do for the dawning "age of transcontinental travel" and the economic growth it could make possible.[16]

Despite the costs involved, Nixon made clear, revamping the nation's outmoded highways was the path opposite the "back road to socialism." The federal government had recently appropriated $500 million for highway spending. But this, Nixon argued, failed to scratch the surface. Progress demanded far more. He called for a $50 billion spending package that, over a ten-year construction period, might make a "good start" on reconfiguring the nation's highway infrastructure. The puzzle for Nixon lay in justifying such action within the Eisenhower administration's conservative political framework. Nixon warned against creating "a blueprint for a regimented economy," yet maintained that the nation needed a centralized "vision" and "comprehensive plans" for its economic future.[17]

As if trying to obscure the incongruence, Nixon deflected attention from the scale of the proposed federal outlay and instead emphasized the functions, both symbolic and real, of such spending. National highway development could be immutably capitalist and decidedly American, refuting communist ideology in political and economic terms. With the right choices, Nixon claimed, a national highway system would reflect an "American philosophy of government . . . rooted in individual rights and obligations." A federal highway system would be not an emblem of the kind of statism associated with socialism and communism, but rather a tool for the construction of a contrasting political economy—and a weapon in America's Cold War arsenal. Gridding out the United States from border to border and coast to coast, a new highway system would be a paragon of American-style state building, one that might well thwart "the communist conspiracy" supposedly intent on destroying the United States through "internal subversion." An outlay of $50 billion, Nixon averred, was a cheap price to pay to "meet that threat," if their design demonstrated "the superiority of our form of government"—and especially if their infrastructural outcomes helped produce a national economy that was not "fat and static," but rather "dynamic and expanding."[18]

Even as Nixon indulged in Cold War abstractions, he revealed that the manner in which the federal government financed and constructed the interstate highway system mattered. His dramatization of statism and its dangers verged on hubris. But it allowed him to present the beginnings of a plan for state building that could mobilize support for a project that Republicans might otherwise see as akin to the very governmental practices to which they objected. The key was to build fiscal infrastructure that diverged from the "excessive taxation" conservatives assailed and converged instead with the spirit of "free enterprise" that had long marked their attacks on the New Deal and its legacy.[19]

For the Eisenhower administration, the interstate highway system's development hinged on the design of a funding mechanism that was acceptable to conservatives. This, in turn hinged on the men put in charge of designing and operating that mechanism. Only those who embodied the notion of free enterprise would do. The Eisenhower administration set about selecting the best men it could find from the world of business to transform this vision into reality in the shape of the interstate highway system.[20]

After Nixon's address to the governors conference in 1954, the Eisenhower administration formed the President's Advisory Committee on a National Highway Program, known as the Clay Committee after its chairman, Eisenhower's longtime confidant Lucius Clay. Clay selected his committee men "from private business." Stephen Bechtel, president of the Bechtel engineering and construction company, S. Sloan Colt, president of the Bankers Trust company, and William Roberts, president of the Allis-Chalmers manufacturing company, guided the planning process. Labor had some representation too, with David Beck, president of the International Brotherhood of Teamsters, joining in. Observers did not have to think very hard to realize how highway development and business opportunities entwined on the committee. Teamster truckers could travel highways built by large engineering and construction firms like Bechtel's with the help of Allis-Chalmers's earthmoving equipment. And alongside these on-the-ground interests, finance capitalism received considerable representation, too: Besides Colt, whose bank traded heavily in government securities, Clay sat on the board of the Marine Midland Trust Company, and Bechtel on the board of J. P. Morgan and Company. These men knew how profitable bond-based public works projects could be for lending institutions.[21]

If profit motivated these newly minted state builders, that was a good thing—from the Eisenhower administration's perspective, at least. And the administration made clear that these new state builders were fully in charge. When Eisenhower also formed an "interagency committee" composed of government officials to study highway planning, its members were told they served only "in an advisory capacity to the Clay Committee."[22] The empowerment of business leaders as government officials also extended to Francis du Pont, scion of the family whose vast business interests had long spilled over into conservative activism. Brought into the Eisenhower administration as a political favor following his prolific campaign fundraising, du Pont oversaw interstate development first as commissioner of the BPR and then as a special assistant to Secretary of Commerce Sinclair Weeks.[23]

Like the other business leaders who became state builders, du Pont represented the change he wished to see in government. He found it "most gratifying" to observe "the difference in the caliber of men" Eisenhower appointed.[24] "New Deal 'do-gooders'" were out; businessmen were in.[25] Du Pont also embodied the paradoxes that shaped the interstate highway project. As a conservative fundraiser and activist, he had long worked to unravel the forms of taxation, governance, and state building associated with the New Deal. But in his new role as a planner of federal infrastructure he had to navigate the competing impulses that filled the Eisenhower administration as its officials sought how to deliver a national infrastructure project while avoiding detested economic and administrative practices.

The Clay Committee led the search for capital to fund the interstate project. It held hearings in which it asked twenty-one highway-related organizations for their opinions on the interstate system and their preferred methods of financing its construction. The hearings began on October 7, 1954, a few weeks before Democrats took both chambers of Congress in the midterm elections. Each group that testified supported improved national highways, but there was little agreement about how to pay for them. Some favored direct application of tax revenue via pay-as-you-pave financing, while others called for ending all direct federal taxation and replacing it with state taxes or borrowed money. The little agreement that did exist amounted to a general aversion to raising taxes. It came as little surprise when the Clay Committee decided that bonded debt offered the clearest path forward. Bonds could deliver a large pool of money up front for the interstate project and avoid the need to raise taxes or impose new ones to cover costs. Although bonds burdened taxpayers with debt service payments, the cost of this interest remained out of sight, felt in the pockets of all taxpayers but seen only by those who knew where to look for it in federal statutes that guided borrowing arrangements.[26]

The Clay Committee finalized its financing plans and completed its policy proposal, *A Ten-Year National Highway Program: A Report to the President*, just as the Democratic congressional majorities took office in 1955. The committee's report called for completion of the interstate highway system over the next ten years at a cost of $27 billion. It recommended that the federal government provide $25 billion of this total, with $2 billion falling to the states. The committee offered no suggestions for how states should raise their share. And when it came to financing and facilitating the federal share, the committee called for a "Federal Highway Corporation" to facilitate the federal government's fiscal contributions.[27]

This corporation would sell over $20 billion in bonds; when banks bought these bonds, they would deliver capital "sufficient to meet [the federal] share of the costs to complete the interstate system." Revenue drawn from existing federal taxes on gasoline and oil would "be pledged in the first instance for payment of the interest and principal" on these bonds, and any excess tax revenues would be applied to interstate construction to make up the remainder of the total $25 billion federal contribution. Seeking to build every advantage into this new bond market, the committee advised that the highway corporation should hold "a mandatory call" on the United States Treasury "for loans up to . . . $5 billion outstanding at any given time, to assure investors of ability to meet obligations." Banks could rest assured that their investment was safe. Should excise tax revenue ever fail to meet bond payments, a bailout drawn from general federal tax revenue would make up any losses. And lenders would do well: The committee proposed an interest rate fixed at 3 percent amortized over thirty years, yielding a profit of $11.5 billion.[28]

The bond plan had the advantage of delivering capital quickly, but controversy arose over how it mixed tax collection and expenditure with a profit-producing obligation to private lenders. The bond plan made it taxpayers' responsibility to pay not just for the interstate highway project but also that $11.5 billion in interest—money that went nowhere near actual highway construction. Why use tax dollars to repay bonds? Why not use taxes to pay directly for construction? As one reporter noted, bonds "cost a lot of interest money that ought to go into the building of roads instead of into bankers' repositories."[29] Opposition to the bond plan formed quickly along partisan lines, with newspapers reporting that congressional Democrats "had vetoed [it] in advance."[30] Eisenhower attempted to play the part of a moderate realist and told the press that "there was nothing partisan about road building," though the need for the assertion suggested its inaccuracy.[31] Attempting to short-circuit Democratic opposition, Eisenhower and Clay hosted a bipartisan summit at the White House, ostensibly to hear contrasting proposals. Journalists, however, discovered that "no major changes would be permitted in the [highway legislation]" before Eisenhower forwarded it to Congress."[32]

Still, Democrats used the summit to call attention to their issues. The bond proposal, they argued, was costly and appeared to be "financed outside the [federal] budget" by using the proposed Federal Highway Corporation to remove bonded debt from congressional oversight and national-debt accounting.[33] In an era obsessed with the national debt—which Eisenhower had pledged to reduce—this was a concerning analysis and a shrewd political play. Eisenhower tried to claim that the bonds "would be a debt," but not "part of the national debt," but how could this be? "An independent government agency is still a government agency," pointed out Senator Harry Byrd (D-VA), the staunch segregationist and chairman of the Senate Finance Committee.[34] And security for the bonded debt was tax money, like all the other money tallied up in the national debt. Senator Albert Gore Sr. (D-TN), chairman of the Senate Public Works Subcommittee on Roads, wondered how the Eisenhower administration could get away with "spending the money as proposed and say it was not increasing the debt."[35] Another congressional Democrat called the bond plan "sleight-of-hand financing."[36] Senator Dennis Chávez (D-NM), chairman of the Senate Committee on Public Works, made it clear that his party "prefer[ed] to build roads with the interest money."[37]

At the conclusion of the White House summit, a Republican Party spokesman noted that Eisenhower, Clay, and their Republican allies had "failed to shake strong Democratic opposition," an admission he followed, somewhat absurdly, by stating that the summit had nonetheless been "harmonious and non-political."[38] With such pronouncements, the Eisenhower administration kindled the myth of consensus that has long accompanied the interstate story—and 1950s politics in general.[39] Eisenhower forwarded the Clay Committee's proposal to Congress on February 22, 1955.[40] The Eisenhower administration was ready to undertake a record-setting public works project, and it aimed to do so without raising taxes. The functional capacity of modern

FIGURE 5. Dwight D. Eisenhower, seated, poses with the Clay Committee, 1955. From left to right, the committee members are Lucius Clay, Francis Turner, Stephen Bechtel, S. Sloan Colt, William Roberts, and David Beck. Courtesy of the Dwight D. Eisenhower Presidential Library and Museum.

Republicanism—promising state building despite conservative fears of government control—was on the line. According to Chávez, however, the Clay Committee's plan was "so full of holes it would sink" as soon as it reached the open waters of congressional debate.[41]

*

Senator Chávez, despite his misgivings, cosponsored the Clay Committee's proposal as S. 1160 in the Senate, where it became known as the "administration bill." His cosponsorship signaled genuine bipartisan interest in interstate highway development. But newspapers pointed out that Chávez was "already on record as opposing the bond issue feature." This was the crux of the matter confronting Congress and, harking back to the Clay Committee's hearings, the nation at large: Nearly everyone wanted the interstate highway system, but few could agree on how to finance it. The inseparability of financing from public works construction and the vast cost of the interstate highway project meant that disagreement over capital appropriation far outweighed more abstract shared aspirations for the interstate system. In Congress, the bill had to pass through hearings held by Gore's Senate Subcommittee on Roads, which fell under the auspices of Chávez's Senate Committee on Public Works. Gore, for his part, could not "conceive that Congress [would] enact such a measure" as the bond financing proposal.[42]

Administration officials cycled through the congressional hearings, building their case. They had to substantiate two central claims. First, that bonded debt was its own conceptual and functional entity, separate from accounting procedures that measured the total national debt. Second, that such bonded debt was worth its cost in interest payments. The two claims together amounted to the administration's vision of market-based social provisioning.

Francis du Pont, now the secretary of commerce's special assistant for highway development, testified plainly on the fiscal procedures the administration bill outlined. Du Pont had built his bona fides from the 1930s onward with predictable claims that New Deal policies led to "socialism" via "deficit financing."[43] Now, he claimed the Democratic Party was "purely interested in perpetuating its control of government in order to seize our economic machine and the production thereof."[44] The stakes were high: The Republican Party, he said, "must regain control" if Americans were to "remain a free people." Otherwise, "The livelihood and liberty of all citizens will be taxed into disappearance."[45] The overarching goal of the Clay Committee's plan and the administration bill, du Pont made clear, was to "issue bonds that would not be within the statutory debt limit" set by Congress. In other words, the first purpose of the bonds was to separate the cost of the interstate highway project from the national debt and place it instead within the parameters of market exchange, which he and his allies presented as a distinct and freedom-filled space.[46]

The suspicious procedure du Pont outlined looked like an illegal evasion of national-debt accounting. Yet Attorney General Herbert Brownell gave his "wholehearted approval" to the plan.[47] In his view there was nothing to hide: The bonds were detached from the national debt, and this was no cause for concern. But why was the Eisenhower administration following this economic path? Lucius Clay told Congress that his committee had almost not even considered "deficit financing," by which he meant using taxes to pay directly for highway construction. Clay explained that bonds offered "a more conservative way" of raising capital and producing infrastructure. Bonds operated on the terms of capitalistic exchange, and Clay told Congress that his committee had "complete confidence in the ability of the market" to carry out the tasks of state building.[48] The message was clear: The marketplace should govern funding for public works.

Submerging public works in the marketplace helped Eisenhower's officials elaborate their theory of national-debt avoidance. National debt, Secretary of Commerce Weeks told the Senate, accounted for the relation between tax revenue and expenditure. The Clay Committee's bond proposal involved the movement of tax dollars in the form of debt security. But the profit incentives that bonds contained altered the nature of the credit and debt in question. The flow of capital produced by bonds, Weeks argued, amounted to a relation between investors and their expectations: "You buy a bond and you are put on notice when you buy the bond that you can expect the revenue from the . . . gas tax to pay the interest on the bond and to amortize it."[49] Bonds yielded profit in the form of interest payments; direct taxes did not. Bonded debt, therefore, replaced taxation with profit-motivated transactions between buyers and sellers. This, Weeks argued, rendered the tax dollars involved in bonded finance different and separate from those tallied up in the national debt.

If Eisenhower's officials felt confident in their justifications of bonded finance and its removal from national-debt accounting, they still had to offer evidence that there was a market for the debt they proposed—that is, that banks would actually offer to sell capital. On this point, ironically, they snagged themselves on their own claims of national-debt avoidance. The interstate highway bonds, Weeks noted, were not "guaranteed by the full faith and credit of the United States Government." This was one more reason they should "not be listed as part of our national debt." But the absence of a federal guarantee posed problems for marketability, given that tax dollars secured the capital at the core of the proposal. It was not clear how banks could protect themselves against a default if gasoline taxes failed to meet obligations. Weeks answered by pointing to the administration bill's inclusion of a "hedge"—the highway corporation's proposed authority to draw $5 billion bailouts from the treasury. General taxpayers were the lenders of last resort. But why not use general revenue in the first place, or why not use bonds that *were* guaranteed by the full faith and credit of the federal government—in other words, typical treasury bonds? Because, Weeks argued, such plans amounted to "deficit financing," and on this point he steadfastly represented the administration's position: The interstate highways were not to be "thrown into the general program" of taxation tallied up in the national debt like a New Deal program.[50]

The bond proposal was held together by ideological statements and marketplace fictions; it remained to be seen whether that would be enough to convince Congress of its viability. And even if Congress bought into the financial logic, another question remained: Was this move to market-based financing worth the cost? Seeking further explanation for why taxpayers should embrace the $11.5 billion in interest payments, Congress called on George Humphrey, secretary of the treasury. Humphrey deployed the standard administration lines. To begin with, he claimed that the proposed Federal Highway Corporation was designed to pay off the bonds itself, using the phrase "self-sustaining" to describe the corporation's activity. This aided the administration's firm commitment to "a less oppressive tax system." The goal was to build the interstates while "cutting back the volume of Federal expenditures" and "the load of general taxes, which weighs so heavily on the economy." Once again, the magic of bonds was on display. Bond interest, Humphrey averred, was not a burden. It was just the price of moving the statist complex of taxation, revenue, and expenditure into the supposedly freedom-spawning marketplace.[51]

According to Humphrey, the cost of interest pushed the economy onward, carrying particular social arrangements with it. The profit-tethered nature of credit, debt, and interest reflected the American tradition of free enterprise, he asserted. These facets of financial life reflected arguments that underpinned the Cold War and the political economy of the Eisenhower administration. "Instead of living under the orders of a dictatorship," Humphrey explained, "we live under a system of incentives. The incentive is a money incentive." Such an arrangement, in this line of reasoning, produced a crisper system of use and payment for public works than taxation could provide. The hope of profit incentivized lenders to buy bonds. They were spurred to make such purchases with the expectation of interest accruals from interstate users who, apparently voluntarily, made the bonds profitable as they used the highways and paid their excise taxes—actions they, too, were incentivized to take by whatever other profits they chased out along the nation's new highways.[52]

To facilitate this kind of economy, the nature of the state had to change. Humphrey imagined government operating like a business. He aspired to a mode of governance in which there was "no difference" between the activities of the state and "the kind of things that private corporations do," especially regarding borrowed money.[53] Among Humphrey's allies, it was a welcome move. "Amortizing capital," one Republican senator pointed out, was a successful principle "whether used by private enterprise or by . . . government."[54] Debt was profitable for whoever owned it. And the interest that bonds carried, in Humphrey's blunt summation, was "worth it," because it amounted to "payment for the privilege of getting something now in order to have the benefit of its use while you are paying for it." In this sense, he argued, bonds offered a nation-sized instantiation of the "consumer credit" that proliferated in 1950s America.[55]

Yet when it came to infrastructural investment, that debt-based model posed problems. As one congressperson noted, a point might come when not just some but "all of the gas tax" went into paying down debt. If that came to pass, how would government pay for ongoing highway development? Humphrey was relentless: "You have to do that by a similar process to this," he replied. "You will have to raise part of the money and borrow part, and keep doing it." Another congressperson pointed out that over time this arrangement seemed destined to make it "harder . . . to balance the budget, because of all the interest." But this did not appear to trouble Eisenhower's officials, despite their paeans to balanced budgets. If debt begot debt, that simply spurred a financialized and privatized national economy in which exchange among market actors provisioned society with its wants, free from dictatorial taxes.[56]

*

The cost of interest and the social vision surrounding the proposed bond arrangement drew sharp questions from congressional Democrats. They prodded the plan for the highway corporation and the bonds it would issue. Humphrey had detailed the Eisenhower administration's scheme: The Federal Highway Corporation would be a "self-sustaining" entity, assuming debt and paying it off separately from the national debt, while convincing bond buyers that the debt they purchased was a safe investment. Gore, from his position as chairman of the Senate Public Works Subcommittee on Roads, pointed out that, according to the administration bill itself, revenue for bond retirement came "from appropriations from the Treasury." The highway corporation would control no "assets or other income." Never mind the $5 billion treasury hedge: The earmarked gasoline taxes themselves moved through the treasury—a point Weeks had conveniently avoided in his testimony. How could this arrangement be self-sustaining in the manner Humphrey had indicated?[57]

The question produced a semantic battle, as Humphrey explained the relationship the Eisenhower administration envisioned between capital collected from taxpayers and capital borrowed from banks. When pressed, Humphrey conceded that perhaps the proposed highway corporation was not self-sustaining after all: "The word 'liquidating' would be better," he amended. It was "the pledge of the gasoline taxes" that "liquidate[d] these bonds" and separated the highway corporation's activities from the national debt. He then tried to divert attention from the term "appropriation," putting "pledge" in its place. But Gore was unconvinced. The bond retirement was "more than a pledge," he countered. The highway corporation had no capacity to raise its own capital; it depended on "an outright appropriation from the Treasury of the United States enacted by Congress." Humphrey persevered. The administration bill, he said, provided "for the dedication of certain revenues to certain purposes." This sounded like typical general-revenue earmarking, which was doubtless more than a pledge. But Humphrey insisted that it was not—that the administration bill separated the highway corporation and its bonds from general revenue and national debt. Trying to fight off the counterarguments that bonds were part of the national debt, Humphrey strained to replace the material reality of the capital movement in question with fuzzy phrasing. Eventually, he went so far as to claim that the bonds were not a debt at all: They were "an obligation and not a debt within the strict meaning of the word 'debt.'"[58]

Sustained or liquidated, pledged or appropriated, indebted or obligated: Humphrey waged a linguistic campaign that could not, in the end, dodge accounting practices. "I do not want to quibble about words," he eventually declared. In a last effort, he suggested that the congressional hearing should avoid getting into "technical meaning" when it came to bonds and debt. All that mattered was that the highway corporation "contain[ed] its own basis for liquidation," and if Gore disagreed, that was "a difference of opinion," nothing more. That difference of opinion, however, proved fatal to the bill. Hale Boggs (D-LA), chairman of the House Subcommittee on Roads, concluded that the administration bill did not reflect "sound financing." He objected to the changes in government it appeared to propagate. "If you advocated for this plan for highways," he asked, "why would you not likewise . . . create Government corporations to do all kinds of things in the Government and keep that out of the national debt limit and mortgage ourselves forever?" His rhetorical question probed the limits of replacing taxes with debt, a substitution that promised to remap the provision of public works and revise the relations binding citizens, private interests, and government.[59]

Despite Boggs's blunt characterization, the debate had not yet concluded. Among the loudest voices opposing the administration bill, offering something of a capstone to the proceedings, was that of Senator Byrd. Though he was a Democrat, Byrd had a reputation for fiscal conservatism and had long critiqued federal spending. He had broken frequently with his party to vote against New Deal programs, and he opposed the presidential campaigns of Harry Truman and Adlai Stevenson. In contrast, he called Eisenhower's election "the first ray of hope we have had for many years" and favored the new president's talk of fiscal caution and balanced budgets. But he proved to be no more enamored of debt service costs than he was of New Deal financing plans. The Eisenhower administration's approach to interstate highway planning alarmed him. To begin with, he found the interstate project to be nothing less than "the longest step yet taken toward concentrating power in the Federal Government." Handing "control of 40,000 miles of our most important roads" to the central state apparatus would surely raise "the iron hand of the federal bureaucracy" as never before. And no matter what mechanism the federal government chose to fund the massive project, it would require "permanent indefinite appropriations" to sustain its existence, locking in place fiduciary relationships that Byrd intended to weigh very carefully.[60]

While Byrd's ideological qualms were unsurprising, much of his more precise concern rested on how the administration bill's use of bonds subverted democratic oversight of the executive branch's use of public revenue. The problem was the administration bill's plan to remove the Federal Highway Corporation from "annual appropriation control by Congress," a striking disruption of oversight, especially given the size of the federal government's commitment. Byrd reminded Congress that the federal contribution, apparently shorn of congressional management, would come from "a Government Corporation without income or assets." It would "borrow $21 billion . . . without declaring it a debt." As Humphrey's testimony made clear, there was no mathematical formula or material chain of capital movement to explain this; the metaphysical evasion relied on the administration's argument that locating bonded debt in the federal highway corporation rather than in the treasury constituted its separation from the total national debt. "Camouflage it all you please," Byrd told his colleagues: The fact remained that "the bonds issued by this Corporation will be a Federal debt and a general obligation of the Government." The proposed bonded debt warped fiscal democracy by tethering taxpayers to bankers and reducing Congress's oversight role. And in the process, it appeared to remove tax money from standard accounting procedures, distorting the national debt and "creating fiscal confusion and disorder."[61]

The administration bill, Byrd concluded, was "incapable of honest Federal bookkeeping and accounting." Recategorizing debt and hiding it away in a "dummy corporation," the bill amounted to the federal government keeping "a dual set of books." "The least [Congress] can do," Byrd summed up, "is to keep the books honest and make full disclosure of the obligations we are incurring." Lawmakers, Byrd cautioned his colleagues, must not "avoid financial responsibility by legerdemain" nor "evade debt by definition." Byrd's arguments, said one senator, left the administration bill "as dead as Judas Iscariot." On May 25, 1955, the Senate voted down the administration bill, eliminating bonds as a federal capital source for the interstate highway project. The stage was set for excise taxes to pave the way for interstate construction.[62]

*

The financial legislation that finally enabled interstate construction in 1956 bore little resemblance to the proposal the Eisenhower administration had sent to Congress the year before. Democrats set about earmarking new taxes on automotive goods to fund the interstates. Much as Republicans had hoped the interstates might stand as a durable manifestation of their tax-averse state-building methods, Democrats now sought to pose the infrastructure project as their own model of fiduciary ambitions. Some also jumped at the chance to exploit the apparent switch between the two parties. Rep. Wright Patman (D-TX), for instance, said he was "surprised" that Republicans, who were "always talking about fiscal responsibility, sound dollars, sound money" had advocated "the reverse" in the administration bill. He blamed the Clay Committee, which "was weighed down with investment bankers, commercial bankers, brokers of Government bonds." Their "first consideration" was "the bankers and what they get out of it," and their proposal reflected nothing more than the interests of profit seekers "looking for more debt paper."[63]

Republicans responded with derision. "Where have you been for the last 20 years on your fiscal responsibility?" asked Rep. Leon Gavin (R-PA). "You should have been thinking about fiscal responsibility back in the days when we had the NRA, the WPA, the PWA, the youth movements, the ballet dancers, and the goldfish swallowers." Democrats had never "thought of bringing in a companion tax bill to pay for anything," he charged, meaning earmarked excise taxes, "until this highway legislation was introduced." But the logic of Gavin's accusation was flawed. If the New Deal framework had been as fiscally irresponsible as he suggested—and if Democrats should have deployed companion taxes long before—his rejoinder made the case *for* using companion taxes to fund the interstate project. Gavin's comment also ignored the financial design of the Social Security Act, which was where Democrats turned for inspiration as they devised their fiscal plan for the interstate project. They called on none other than Secretary Humphrey, who had argued for the merits of bonded debt and interest, to explain the benefits of the Social Security tax model.[64]

The Senate heard testimony from Humphrey in May 1956, a year after the defeat of the bond proposal. In the interim, as the Eisenhower administration suffered at the hands of the national media for failing to deliver the highway system, Humphrey decided that he would "heartily endorse" the Democrats' "pay-as-you-build program" consisting of excise taxation and direct revenue expenditure.[65] The change could not have been more dramatic: Humphrey departed from every argument he had made the previous year and seemed to jump sides. He agreed with the Democrats and recommended that Congress "follow a practice similar to . . . social security" and collect earmarked taxes from highway users to be "deposited in a special fund."[66] This trust fund would provide the capital required for construction. It was a plan for state building based on a basic relation between taxpayers contributing earmarked excise taxes toward the construction of public works that, in turn, supported the activities from which the excise tax was collected.[67]

Even Senator Byrd, sometimes a critic of earmarking, was convinced—perhaps because he was impatient to get the interstate system underway, already recognizing how it might serve his segregationist agenda as a means of bulldozing and walling off Black communities.[68] He persuaded the Senate Finance Committee he chaired to support the plan. In doing so, he compromised with his own party, leaving his issues with earmarking aside for the chance to have a stronger legislative hand in determining the increase in taxation and expenditure. On April 27, 1956, the House approved the Federal-Aid Highway Act of 1956, including Social Security–modeled excise taxes that funneled into the newly created Highway Trust Fund, by a vote of 388–19. The Senate approved the legislation 89–1 on June 26. With veto-proof majorities secure, the bill reached Eisenhower a few days later.[69]

After Eisenhower signed the act into law on June 29, the federal government prepared to begin collecting revenue to build the interstate system. Fiscal year 1957 began two days later, and at the stroke of midnight, federal excise taxes on gasoline and diesel rose from two to three cents per gallon, excises on rubber tires rose to eight cents per pound, and trucks weighing more than 26,000 pounds began paying $1.50 per 1,000 pounds of gross weight. These taxes covered 90 percent of interstate construction costs, with the remaining 10 percent coming from the states.[70]

This arrangement required time to reach peak performance. Excise taxes needed to add up bit by bit until there was sufficient capital to begin construction. As *Newsday* noted, transportation was not part of the equation yet: Rather, "for the motorist, the first noticeable effect will be the enactment of the tax increase."[71] It was headline news: "Fill Your Gas Tank, Buy That Tire Today Before Tax Goes Up," warned *The Boston Globe*. But as the excise taxes accumulated and construction proceeded—opening ever more highway mileage to drivers—revenue grew in the highway trust fund as Americans bought more gasoline, oil, and tires to sustain their automotive existence. While Secretary Humphrey's talk of a "self-sustaining" highway corporation might have been fanciful with respect to the Eisenhower administration's bond plan, the term looked apt when applied to the Highway Trust Fund. The perpetual fiscal motion it enabled was an almost too efficient tool of state building.[72]

The tax structure contained within the Highway Trust Fund was slated to remain in effect until at least July 1, 1972. Along the way, it would fix a powerful fiduciary cycle in place. The initial financing plan at the core of the 1956 legislation called for a sum of at least $25 billion in federal funds. But the trust fund meant there was no upper limit to revenue collection and expenditure.

FIGURE 6. The Bureau of Public Roads offered a visual portrayal of interstate highway financing, 1966. Courtesy of the National Archives and Records Administration.

As long as motorists went about their business, the tax dollars would keep the cycle spinning, ceaselessly transforming revenue into pavement.[73]

*

Democrats had ensured that the interstate highway project would not join the swelling ranks of post–World War II bond-financed public works projects.[74] The excise taxes undergirding the interstate highway system saw the federal government building itself confidently and visibly with its own money. The interstate highway project thus produced a legible instantiation of modern American state building that was the very antithesis of governmental activity kept "out of sight" with the help of privatized and financialized fiscal choices. The excise tax loop, however, proved to be an imperfect solution to the problems of bonded debt.[75]

The Highway Trust Fund established a circular system of excise taxation—generating excise tax revenue from automobility, transforming it into highway infrastructure, and then accumulating ever more revenue as the pavement and traffic flowed. The BPR celebrated that tax-and-pave loop as a bold policy innovation. But these fiduciary dynamics compelled continual highway construction irrespective of user needs and community concerns. As development ramped up in the 1960s, critics drew attention to the destruction that the interstate system's seemingly endless growth wrought across human and environmental geographies. The fiscal infrastructure undergirding the physical infrastructure insulated destructive outcomes from oversight and reevaluations of spending priorities that might have curtailed the worst consequences and produced a more equitable highway system.[76]

While the public, in 1956, could not yet foresee the full impacts of the financing mechanism, the Highway Trust Fund aroused immediate frustrations. Dispirited by the congressional fight and stuck with new costs, the public expressed little support for *any* functional funding structure. In a Gallup poll that year, a meager 11 percent of respondents supported "higher taxes on gas, oil and tires" to build the interstates—matched by another 11 percent who said they still supported "hav[ing] the government borrow money." While Americans in the 1930s had balked at the idea of toll financing, now 40 percent of respondents preferred "a toll of about one cent per mile," an option long since shown to be insufficient to fund construction, and long since abandoned by policymakers. The popular idea of a national highway system, in the end, had produced a distinctly unpopular financing system that grew more unpopular over time.[77]

The enactment of excise taxes avoided billions of dollars in interest payments but brought other costs, unforeseen even by budget hawks like Senator Byrd. The Highway Trust Fund enabled continual highway development and established a petroleum-industrial complex that pushed construction and expansion ever onward.[78] The circularity of the fiscal and physical infrastructure meant that although congressional Democrats had averted debt service payments, they had created other burdens that were not only financial. The presence of perpetually available capital and the absence of clear user need informing interstate planning called into question how much money the excise tax regime really saved taxpayers. What if officials were building highways no one needed or wanted? Compounding costs—financial and otherwise—soon became plain.

From the moment construction began, the interstate highways played a central role in the postwar "culture of clearance" that brought social and environmental destruction to the nation's landscapes. Such devastation was the price paid for the shopping centers, subdivisions, and suburban sprawl that marked the post–World War II decades. And the interstates made all of this possible, scouring landscapes as disparate as wilderness areas, town parks, and urban neighborhoods to facilitate economic expansion.[79] By the end of the 1960s, the vociferous highway critic Ben Kelley—none other than the BPR's former director of public affairs, radicalized by the shattering effects of interstate construction—wondered whether bond-based financing might have kept the interstate project "more nearly within the limits" of wise development.[80] In his 1971 book *The Pavers and the Paved: The Real Cost of America's Highway Program*, he looked back on the debate over bonds.[81] Congressional Democrats had unleashed an "untested devil," as Kelley put it, in the form of excise taxes.[82] He concluded that bonds, despite their problematic interest costs, might have forced Congress to make more attentive choices about the interstate highway project as it progressed. Kelley's hopes may have been overly optimistic. Still, his commentary suggested just how imperfect the excise taxes were.

This problem guided Kelley's continued opposition. In a 1971 article, he and his coauthor Richard Hébert used a counterfactual to explore the startling realities set in perpetual motion by the Highway Trust Fund: What if Congress had added the interstate highway program to the national budget, rather than earmarking its tax revenue? Imagine, that is, if "interstate construction were being paid for today, like everything else, out of the general treasury," and was subject to "periodic review." Under such an arrangement, "no responsible federal official would consider paying for transportation projects in any way different" as time went on. The Highway Trust Fund would be inconceivable. Kelley and Hébert could not prove their counterfactual hypothesis, of course. Still, their analysis made clear that the existing framework for interstate highway financing and construction was at odds with the needs of citizens and communities.[83]

*

Government trust funds produce resilient forms of state building. The originators of the Highway Trust Fund knew this, like those of the Social Security Act before them. The Social Security Act and the Federal-Aid Highway Act of 1956 each produced public goods whose legislative and material designs guarded against change and reform. Social Security has endured far better than many of the other welfare policies implemented alongside it in the heart of the New Deal. Scholars have evaluated the substantive success of Social Security with regard to who is included within and excluded from its social safety net. This has led some historians to focus on the origins of the Social Security Act, emphasizing the political forces that, for example, prevented the inclusion of federal health insurance in the act's provisions. Exploring such failures guides much of the literature. Yet, as historian Linda Gordon writes, deeper insight comes not from thinking about "more or less welfare spending," but rather from thinking about "the shape and spirit of public provision." In other words, some insight can be gained from analyzing paths not taken at the time of the Social Security Act's enactment; still more insight can be gained from investigating why the fiscal mechanics of public policy so often foreclose those paths from ever being explored in the future. Social Security and the interstate highway system offer views of how nation-shaping policies are produced and perpetuated. Like Social Security, the interstate highway system is diagnostic of deeper government practices, illuminating how a governing regime embedded its priorities, for better and for worse, across space and time with dramatic durability.[84]

The Highway Trust Fund—the fiscal infrastructure that enabled the interstate highway system's dual processes of construction and destruction—offered little opportunity for reappraisals of development and its consequences, reappraisals that might have reflected emerging economic, social, and environmental concerns. As interstate highway construction began, a set of pitfalls in modern American political economy and state building came on display. Bonded debt and excise taxation each in their own ways called attention to the manner in which such fraught financial mechanisms of state building could misalign the relationship between officials and citizens, tying the two together yet splitting their interests apart, while incurring a vast range of social and environmental costs. Keeping this unwieldy ship of state on course was a complex task for local, state, and federal administrators alike. By 1956, they had control of a powerful flow of capital. The next task was to devise cultural strategies and technical procedures with which to keep themselves allied and the public aligned in the work of infrastructural state building made possible by the endless funding stream.

3
Men Working

The Highway Trust Fund sprang from the Federal-Aid Highway Act of 1956 as one of the most powerful and innovative state-building tools yet devised in American history. Some commentators, however, wondered if the infrastructure project it was designed to enable would commence as smoothly as federal officials hoped. In September 1956—just two months after capital began to flow to interstate construction—journalist Phillip Yeager reported that "the problem of intergovernmental relations" was "the most severe that the new highway program faces." He worried that subnational authorities might fail to "respond" to the interstate highway legislation. His headline blared the worst-case outcome: "Your State May Block New Roads."[1] Although Congress had steadily centralized decision-making over the years—stretching back to the first tentative steps taken with the Good Roads Act of 1916—American federalism's inherent distribution of power meant constant opportunities for conflict among local, state, and federal officials. How well such officials worked together on the nation's beacon infrastructure project remained to be seen. Formidable fiscal infrastructure was in place, piling up excise tax dollars for use. But the cultural and social infrastructure necessary to transform those tax dollars into a centrally coordinated, nation-spanning development project was less certain.

The governmental complexity of the task presented one challenge; another was the complicated politics of the Cold War, pressing constantly on the array of officials who had to work together in order to build the highway system. Americans were supposed to be suspicious, at best, of large-scale government projects and central authority. Highway administrators at all levels of government would have to find a way around that. At the same time, they would need to fashion bonds among themselves tight enough to fuse federalism's layers of authority together. A shared culture was crucial, one that held them together and also pulled society at large into believing in a shared national endeavor. It seemed vital to craft freedom-filled storylines that differentiated the interstate project from social-engineering projects taking place elsewhere in the world under the specter of communism. Officials did not hesitate to invoke myths of American pioneers, ideals of postwar masculinity and family life, and principles of democracy and free enterprise as they shaped political narratives and institutional arrangements. This was pivotal work for the highway men.

Two decades earlier, H. S. Fairbank had warned that "the partitionment of administrative responsibility between the Federal, State, and local governments" might throw a wrench in a national highway development plan. In the midst of the New Deal's turn to federal planning, he had hoped that the "federal assumption of outright responsibility for a Federal system of highways" would be possible.[2] Yet even then the challenge was that American government—and political culture broadly—was often resistant to centralized authority. Now Fairbank's aspiration would be put to the test. The funding arrangement helped. Needing to contribute just 10 percent of their own funds to interstate highway development, states and localities might be eager to fall in line in return for federal dollars, even if their own routing or construction preferences were different than those planned by the BPR.

But astute observers were not convinced that the financing alone would be sufficient to unite local, state, and federal highway officials. In November 1956, the Tax Institute of America convened a conference to evaluate challenges its members foresaw. The Tax Institute was a prominent organization, taking taxation as a broad subject of inquiry that included questions about governmental capacity, social outcomes, and the physical work of statemaking. The organization published research on various public policy areas and held an annual conference on a particular topic of lawmaking. The interstate highway project, fundamentally a product of tax policy, was an obvious choice for the 1956 conference. Gathering at a hotel in Princeton, New Jersey, policy analysts, political scientists, and economists shared papers that ranged from optimistic predictions of long-term economic benefits to more skeptical appraisals of the likelihood of the interstate system overtaxing American society. Nearly everyone agreed that a highly cooperative instantiation of federalism would be required to make the interstate highway system a reality. The conference's chairman, Homer Scace, a research associate in the New York State Legislature, identified the central point of convergence: Who in government would "be held responsible for getting the job done?"[3]

Edward Lutz, a professor of public administration at Cornell University, explored the issue in his presentation. He identified a need to devise administrative practices capable of governing the project's construction and ensuing existence. The Federal-Aid Highway Act of 1956 had "emphatically" failed to structure "an enduring pattern of federal-state relations" that would be necessary, in Lutz's judgment, to "confidently plan the long future" as the interstate system's administrators intended. The act reflected a clear move toward consolidation of political management and economic planning. Yet it had only "intensified" the "issues of federal-state relations" that defined American politics and often obstructed forms of centralization.[4] Frank Moore, president of the Government Affairs Foundation, shared a similar concern that progress toward the construction—much less completion—of the interstate highway system depended not on the availability of capital, raw materials, machines, or labor, but rather on "the extent and pace at which we move toward solution of the problems of intergovernmental relationships."[5] If the officials scattered through federalism's layers could find unity among themselves, they would be able to move forward together and craft a new infrastructural future.

*

Cold War politics made federal funding a complicated source of unity among officials, at least in public. The interstate highway project signaled the American state's interest in organizing masses of people and capital through the alteration of the environment in pursuit of economic growth. This was a standard goal for postwar nations moving in the current of high modernist ideology, irrespective of capitalist or communist allegiances. Yet in the flux of the Cold War—bound to the manufactured necessity of contrasting the United States with the Soviet Union—proponents of federal development projects had to claim that the activities they prescribed, funded, and directed were the product of free will, free enterprise, and other atomistic and unbound forces supposedly opposite to Soviet collectivism and control.[6] And few embodied the contradictions of Cold War American statism more clearly than highway officials. They were on the front lines of the largest centrally planned public works project in American history, and they were also under pressure to close ranks within the structure of federalism and consolidate their distributed local, state, and federal powers. The promises of the interstate highway system might be sufficient to motivate just such a collective effort. Even so, they would need to explain their behavior not only to the public but to themselves.

An important task in this effort was the construction of what the political theorist James Scott calls the "public transcript" of government action. The public transcript is the official story of power that justifies authority. While the public transcript must appeal in some measure to common interests, it is also, crucially, "the self-portrait of dominant elites," designed "to affirm and naturalize" their plans to themselves as well as others. Its production is part of the "maintenance work" of government, a performance of surety and self-confidence in the matters at hand. Citizens are an important audience, but so too are officials themselves, who become "consumers of their own performance," rendering their own narrative "something like a self-fulfilling prophecy" as they bind themselves in their own pronouncements and agendas. The public transcript, in the end, dispatches incongruities and insulates officials from political dissent of either grassroots or formal varieties. In naturalizing the decisions of power wielders, it conceals the "dirty linen" of state building.[7]

The public transcript of the interstate highway project emanated from the halls of central government and aimed outward and inward simultaneously. It had first to contend with the vexations of Cold War politics. The challenge was to explain what made a federally funded, centrally planned infrastructure project distinctively American, aligned with the ideals of freedom and free enterprise. As early as 1950, George Romney, a leader of the Automobile Manufacturers Association and later chief executive of the American Motors Corporation, told a symposium of officials and public policy experts that American highway development was "unique." It had no analogue in Europe, he claimed, much less the Soviet Union. The United States was not run by "master-planning." The American highway, "like the vehicle that created it," was "the product of the people, a thing made by the people for the people."[8] Romney's statement tracked with a broader approach to describing American economic life as the 1950s wore on—as one advisor soon reminded Dwight Eisenhower, it was important to depict the nation's political economy as "an overall system of individual endeavor."[9]

How could that image be harmonized with the centralized development of automotive infrastructure? A telling answer came from the Ford Motor Company's *Freedom of the American Road*, a combined book and film project aimed at drumming up public and official support for the interstate highway project. Ford's interest in this boosterism was plain: More and better roads would help car sales. But the company's leaders played all the right notes as they contemplated massive state activity. Henry Ford II, president and CEO, wrote in the book's opening pages that federal highway planning was nothing more than a response to "individual citizens" yearning for more "elbow room"—for "freedom to come and go as we please in this big country of ours." The interstate highway project was nothing like the high modernist infrastructure projects mandated by central authorities out in the communist sphere of influence. The American effort was simply a result of "the private citizen working in joint voluntary effort with others like himself"; it was "evidence of democracy in action." Ford assured readers that they would find "no master plan" for action as they read on, although a map of the interstate highway system's as yet unbuilt routes followed on the next page.[10]

As officials and boosters strained to hold conflicting ideas together, they called on luminaries of American mythmaking to build their case. Also in *Freedom of the American Road* was an essay by Bernard DeVoto, the Pulitzer Prize–winning historian of the American West. DeVoto identified the aspirations that had long animated federal highway officials. Sounding like H. S. Fairbank, he wrote that the changing needs of the "American economic system" were the "primary reason" the nation's existing highways had become "inadequate." Even an anodyne statement such as this verged on dangerous national planning talk in the mid-1950s, but officials had spent years building this case prior to the emergence of new Cold War rhetorical demands. Perhaps DeVoto realized it was foolish to attempt to ignore that older argument. He stated that it was time to stop building highways "piecemeal," and instead to ensure that automobility could progress in a "coordinated" fashion across the nation and into its future. Even the most advanced stretches of highway development—DeVoto pointed to the New Jersey Turnpike, the New York State Thruway, and the Pennsylvania Turnpike—were premised on economic life in individual states. This scale of thinking and building might have been fine for an era of "horse-drawn traffic," but it could not hope to meet the "needs of the future" as the nation and its economy grew.[11]

DeVoto played a more subtle game than Romney or Ford had as he worked back toward solid Cold War footing. Americans, he averred, were not falling into statism so much as perfecting the very frontier impulses that historians like him claimed were responsible for American exceptionalism. Those impulses were as old as colonization itself. "Maybe it began as a footpath through the woods to the meeting house," he wrote. "A narrower path from the saw-pit entered it, and the trail which the traders took toward the notch in the hills." It was a vision of rugged individualism transforming into harmonious and productive development, an image recognizable to anyone steeped in the American mythos. A couple of generations had passed since Frederick Jackson Turner, a still more famous historian of the American West, had delivered "The Significance of the Frontier in American History," his 1893 address to the American Historical Association, describing the final conquest of the furthest boundaries of North America. Conquering the frontier, in Turner's telling, had produced an exceptional nation—one marked by individualism above all. Now, DeVoto asked an implicit question: Could the colonization of the American frontier and ensuing economic expansion have happened without good roads? "There was never a time when the road was not of absolute importance to the individual and to the community he lived in," he answered.[12] The interstate highway system was heir to the very practices that had made the nation all that it was.

One connection to the American pioneering spirit that came with an element of truth was the racist legacy of settler colonialism. If "manifest destiny" had once been the language of violent geographic revision, the postwar period saw the emergence of phrases like "slum clearance" and "urban renewal" taking aim at "blight."[13] This veiling language, draped atop the destruction of Black neighborhoods, had grown more explicit over time. The BPR's *Toll Roads and Free Roads* report of 1939 had noted that "the interior of the city is ripe for . . . major change." Five years later, the *Interregional Highways* report that laid out the basic plan for the interstate system noted that construction would require "razing" "large tracts of blighted urban lands," which the report's authors "regarded as an end desirable in itself."[14] By 1958, the American Road Builders Association (ARBA)—an influential group whose members tended to move seamlessly back and forth between government and industry—called the interstate highway system a "golden opportunity" to use "highway construction in eliminating blight and slums."[15]

By this time, the director of the American Association of State Highway Officials (AASHO), Alfred Johnson, reported agreement among officials that the interstate highway project "would give them a good excuse to get rid of the local 'n——town,' " and they began drawing up maps accordingly. When those maps turned into pavement, the interstates joined a wave of similarly intentioned infrastructure projects that destroyed as many as 1,600 poor and nonwhite communities across the nation.[16] The legacy of the frontier did good work for the public transcript of power and the white, elite priorities it insulated. In advertising, music, film, and literature, officials and business leaders told Americans that they were most free when traveling on the road—though that freedom, evidently, was not shared in by all.

*

Linking the interstate highway project to the American frontier myth helped highway officials deflect the political problem of central planning. Talk of "pioneers" proliferated; one 1957 *Time* magazine feature on highway development, for example, described how the "panorama of road builders stringing highways across the land reflects a peculiarly American genius, one that lies deep in the traditional pioneering instincts of the nation."[17] Meanwhile, racist aims offered officials common ground on which to gather from their different corners of federalism. Talk of rugged frontiersmen also helped them deploy gender to further differentiate American statism from its Soviet parallels. The era's political culture saw a particular masculinity take shape, one dead set against Cold War anxieties over the supposed "feminization" of American culture, which some claimed was susceptible to communist "penetration."[18] Highway officials—who were nothing if not agents of the central state—ramped up their reliance on gendered tropes and projections as they explained, to themselves and others, the work that they were doing.

In this vein, the Eisenhower administration's first public plan for interstate highway development articulated a vision in which "the breadwinner uses an automobile to get to work; the housewife to shop."[19] And while the administration was busy describing the interstate highway project in relation to breadwinners and housewives, it also formed the "Manpower Commission," which published six lengthy volumes on national labor capacity while mentioning women in only one.[20] This "gendered imagination," as historian Alice Kessler-Harris put it, defined cultural and political work and shaped economic plans and the administrative apparatuses necessary to make those plans reality.[21]

The interstate highway project fit the contours of this sex-differentiated economic landscape. As interstate construction began in 1956, 22,000 American men earned engineering degrees, alongside 62 women. The BPR does not appear to have employed any of these women until a single female engineer appeared on the payroll in the early 1960s. The college training grounds for those attached to the interstate project were almost exclusively male. The pattern was part of the postwar reversion to sexist employment practices. While the war years had seen the number of women employed as engineers increase by a factor of ten, those numbers dropped after 1944. Colleges and universities played a role in the decrease. The Massachusetts Institute of Technology, which began admitting women in 1882, concluded in a 1956 study that "women were not successful undergraduates" and came close to ending coeducation; the campus climate was presumably a less than welcoming or equitable place for aspiring women engineers. Rensselaer Polytechnic Institute, another key mid-century training ground for the engineering profession, graduated its first female students in 1945 as part of an accelerated wartime program. By 1958, however, the school's leaders were encouraging faculty and administrators not to "stress the fact" that women could apply to the school—much less attend—because "we don't want too many gals around."[22]

The gendered dynamics did not stop there. Labor was sex differentiated, too. Interstate highway construction amplified labor and commodity markets spanning from oil drilling to heavy equipment production to the work of laying down pavement.[23] All of this fit within an economic policy regime that, in historian Linda Gordon's words, contained "a silent modifier—'male.'"[24] The interstate system benefited male labor, whether in the form of workers drilling oil and running bulldozers or commuters zipping in and out of new suburbs. Women, in other words, had less access to the taxation-derived benefits of the capital circulated, redistributed, and invested via the interstate highway project. One critic soon argued that the interstate highways instead left women "stuck" in the cul-de-sac landscapes of the 1960s, where Betty Friedan identified "the problem that ha[d] no name."[25]

Other problems filled the gendered geography of development and economic growth. The postwar years saw a sustained effort to code car culture and the act of driving as masculine. There were at least three interrelated factors at the center of the sexist turn in car culture after World War II. First, the needs of wartime had put more women than ever behind the steering wheel, whether in domestic or commercial capacities. Second, just as women's participation in the labor market and civic life was curtailed following World War II, the simultaneous proliferation of car-dependent suburbs meant that the number of women driving (by necessity and to fulfill the gendered demands of homemaking) continued to increase. Third, the era's interest in statistical science and the presence of ever more finely tuned insurance mechanisms meant there was ready and compelling evidence that women were, in fact, better and safer drivers than men. More women were driving than ever before—and more skillfully than men.[26]

One response, as historian Roland Marchand has found, was the assertion of "a compensating image of masculinity."[27] Attempts to establish the gendered hierarchy of the roadway abounded. Take M. M. Musselman's popular *Get a Horse! The Story of the Automobile in America*, which appeared in 1950. It recognized the rise of women drivers and stated that "when it no longer took muscle to start the family car the ladies quickly commandeered the family bus and the era of the accordion fender and the baffled traffic cop was at hand."[28] Perhaps spurred by heightened fragility, men like Musselman worked to assert sexist stereotypes and, when possible, tie them to the American frontier mythos of rugged male individualism. It took little time for psychologists to suggest that men's collectively worse driving record was proof that they were strong-willed, daring, and suspicious of authority—key positive markers of the ideal Cold War American man, juxtaposed with the supposed characteristics of his Soviet counterpart.[29] Jack Kerouac's *On the Road*, appearing in 1957, made male unruliness inseparable from the automotive mise-en-scène—cars and highways were the virile antidote to the stifled, feminized conformity of the various communities from which Kerouac's autobiographical character and his pals continually fled.[30]

Advertisers—newly powerful figures in the 1950s and 1960s—did their part to strengthen the bonds of automotive masculinity. Because women drove ever more frequently, they were necessary targets of advertising. But the total picture of salesmanship was one in which real *drivers* were men, and only with shiny new cars could they conquer new frontiers. "He wanted a Fiat Spider. She wanted a Fiat '600," read one advertisement for a pair of new releases. The Spider was "all the car a man could want," full of "plenty of zip." In contrast, the '600 was "perfect for scooting to the supermarket."[31] If this played up the era's interest in breadwinners and homemakers, other ads moved toward the competing culture of sexual liberalization, although in no less sexist terms. Ads for the new 1961 Plymouth four-door, featuring a spacious backseat, portrayed a young woman touching up her lipstick and a straightforward enticement: "The engine is idling, the radio and heater are going . . ."[32]

Rugged frontiers may have become smooth highways, but in all its expressions, masculinity helped American men tell stories about their mythic pasts and present identities that appeared to resolve some of the paradoxes of American statism during the Cold War. The gendered narrative helped forge an administrative fraternity that bound together local, state, and federal highway officials. Notions of freedom and movement on the one hand, and mastery and domination on the other, defined the culture under construction.

*

A key site of cultural formation for highway officials was the annual AASHO meeting. Men from all levels of federalism bonded, discussed policy, and strove to guide their nation-spanning paving project. The officials—especially those in leadership positions—choreographed masculine informality and created a clubby atmosphere in which insider status was cemented. Entitlement was cheered on, even as strict adherence to the central plan was policed. As one official remarked, the policy meetings were a setting in which "empire building is encouraged and bureaucratic regulations and controls are being promulgated."[33] There was no tolerance for deviation, even as apparent frankness and honesty characterized the public transcript at the meetings. Presentation titles such as "Random Uncensored Observations of a New Highway Director" abounded, though the contents of the presentations were anything but random.[34]

But the performances on display projected a particular fantasy of maverick male freedom. The men built a space for themselves in which Ellis Armstrong, a longtime highway official who served as commissioner of the BPR in the early 1960s, could brag about enjoying the "oscillation" of "burlesque dancers" in the company of the secretary of commerce. Doubling down, he told his colleagues that he found inspiration to do his work as a highway official "every time [he] watch[ed] a luscious, tantalizing example of female pulchritude go tripping down the street."[35] Even lingering discomfort with centralized administration was expressed in a similar manner. One state-level official discussing the complexity of coordinating local and federal authority said he had "as many bosses as the oriental potentate who still practices bigamy." He was not complaining, however, but rather describing how much roadbuilding action was underway. The Tax Institute analysts who had discussed the delicacy of "intergovernmental relationships" a few years earlier need not have worried; embrace bigamy, and everything would be just fine.[36]

With structural and cultural factors increasingly aligned, centralized government pushed interstate highway construction ahead with rising confidence and efficiency. The task moved forward with the leadership of a wider range of actors than had previously been involved with road work. Highway engineers were now joined by economists, business leaders, politicians, military officials, and federal administrators. But engineers, in control of the technical and physical design of durable roads suited to evolving traffic needs, held a keystone position. They had long played important roles in shaping legislation, and by the 1950s they held increasing power in agency administration. Everything from construction practices to intergovernmental cooperation came under their purview. At the meeting point between bureaucrats and builders, and themselves some combination of the two, they were responsible for coordinating the mix of governmental authorities and capacities called forth by the Federal-Aid Highway Act of 1956.[37]

The general location of interstate routes had been mapped out in 1947 and reaffirmed in 1955.[38] The next step was for state highway officials to produce site-specific blueprints for construction and get to work. Their general approach in this phase was to find "the shortest route and the straightest alignment."[39] Called "point to point routing," engineers determined the most direct path between two cities and proceeded accordingly. The BPR aimed to produce road conditions marked by "sweeping curves, easy grades, and long sight distances."[40] The interstate system would not accommodate existing landscape features—whether natural or human-made—but rather bulldoze through them to achieve design and traffic goals. Per mile, the total paved area of an interstate highway would be 2.75 times greater than any previous highway construction.[41]

The Highway Trust Fund's first outlay had come on August 2, 1956, delivered to the Missouri State Highway Commission. It covered the requisite 90 percent of a $1.1 million contract awarded to Koss Construction, a firm based in Iowa. Backed by a team of highway engineers, soil scientists, and hydrologists, machine operators got to work transforming 13.3 miles of famed Route 66 into a section of Interstate 44 in Laclede County, MO. This inaugural contract revealed just how massive and complex the national project would be.[42] With the exception of the military, interstate construction in the 1960s was the nation's largest tax-funded project, even surpassing the era's skyrocketing NASA expenditures. It also generated more tax revenue than any nondefense area of economic production and employed some fourteen million workers in construction and related industries, representing 20 percent of the American labor force. Economists estimated that highway construction and use contributed approximately twenty cents to every dollar tallied in the gross national product.[43]

FIGURE 7. A massive cloverleaf interchange on Interstate 5 takes over Glendale, California, 1957. Courtesy of the Federal Highway Administration.

FIGURE 8. Interstate 75 blazes through woodland northwest of Atlanta, Georgia, 1966. Courtesy of the Federal Highway Administration.

AASHO and the BPR had agreed on building and design standards just two weeks before the Laclede contract commenced. Their specifications stipulated at least four traffic lanes, each at least twelve feet wide. All interstate mileage was to be flat and straight enough to accommodate steady speeds of at least fifty miles per hour. Fourteen feet of overhead clearance (say, when passing under another road) was required, it was claimed, to ensure the movement of military vehicles and weapons. Pavement would be laid on one side of the right-of-way taken for construction, leaving space for the addition or expansion of lanes in the future.[44] About 85 percent of interstate highway mileage was new construction, with the remainder accounted for by upgrades of existing highways. Some earlier state highways—particularly the Pennsylvania Turnpike and the New York State Thruway—had served as inadvertent testing grounds for design standards and construction procedures, and portions of these were absorbed without modification into the interstate system. For new construction, pavement was wider, thicker, and stronger than any previous specifications had called for. On average, it demanded nine times as much excavation, seven times as much tar and asphalt, over four times as

FIGURE 9. The Bureau of Public Roads created a cross section of Interstate 95 for demonstration purposes, 1963. Officials aimed to show off the thickness of interstate highway pavement and the depth of undergirding base material. "The photo provides indisputable evidence that the tax payer is receiving full value for every highway tax dollar," the agency announced. Courtesy of the Federal Highway Administration.

much cement, and over three times as much steel, stone, gravel, sand, and other base materials as older highways had.[45]

It did not take long for construction to reach dramatic proportions. By the 1960s, the BPR began to make public declarations of what appeared to be a decisive policy victory after years of careful groundwork. The agency's attention to narrative-crafting, however, suggested just how tenuous the harmony in highway development was. In 1962, officials pointed their triumphal narrative at the public in *America's Lifelines: Federal Aid for Highways*, a pamphlet the BPR began to publish annually. The title alone helped craft the agency's public transcript of power, as it worked to remap the nation's infrastructure and political economy. Year by year, *America's Lifelines* reminded the public that the largest public works project in American history would be impossible without the BPR's "approval and control, in each succeeding step of the process." The agency's "unifying and coordinating influence" was the prerequisite to infrastructural success and the safeguard against failure.[46]

The BPR, in its telling, was playing a vital role. Describing the interstate system as a pressing necessity rather than a product of subjective priorities, the agency portrayed itself delivering the nation from something like a crisis. Cheap petroleum and the entwined politics of highway development were, as ever, at the heart of the nation's "growing tide of traffic." But the BPR made no mention of these factors and the human choices underlying them—much less their destructive consequences. Rather, traffic was depicted as an inexorably demanding force. Only the interstate highway system would be sufficient to meet the needs of all those spinning wheels—only the interstate system could resuscitate the United States from an ill-planned automotive past and avert a dead-end future. It was an effective way to naturalize government authority and action. It also helped cast American government as exceptional and different from other varieties of central state ambition found in the Cold War world. In the BPR's narrative, aimed at officials as well as the public, the interstate highway project did not engineer society, but rather resolved emergency conditions. Such a framing offered a legitimation of massive state action that could be excused even by those banging the drum of anti-statism. They might even go so far as to claim that the government work underway was a critical means of protecting the individualism that had, supposedly, defined the nation since the time of frontiersmen.[47]

As ever, the rhetorical line was a fine one. *America's Lifelines* told the public not to worry over the looming presence of federal authorities imposing a nationwide development scheme. If Americans wanted to enjoy the "wide range of economic benefits" that would flow from the interstate system, they needed to accept that the system's construction was a "complicated business" whose very purpose was to transgress "state, county, and city lines." But *America's Lifelines* assured readers that subnational officials would have important roles throughout the construction process.[48] Take the BPR at its word, and the project would be a success. Soon—repeating a favored phrase among highway officials—motorists would be able "to drive from coast to coast without encountering a traffic light or stop sign." In short order, the interstate highways would connect "principal metropolitan areas" and "industrial centers," creating a nation-sized web of "production, assembly, and distribution" and "great economic growth." Making such hopeful promises, the BPR told its readers that nothing was more critical than "speed[ing] the job" to completion—a task that required the firm command of federal officials.[49]

*

And the job sped ahead. By 1966, *America's Lifelines* boasted of the 1.8 million acres of land seized for construction and estimated the amount of dirt and rock excavated could "blanket Connecticut knee-deep." The amount of "sand, gravel, and crushed stone" accumulated for roadbed construction was sufficient to encircle the globe in a band fifty feet wide and nine feet high. With the space race front and center in American culture, the BPR announced that the amount of pavement required by the interstate system could provide six sidewalks to the moon. Producing all that pavement had demanded eighteen million tons of coal, thirty million tons of iron ore, and over six tons of limestone, with more on the way.[50] All of this depended on an ample supply of machines and workers. World War II left a heavy imprint here in several different ways. The war had reorganized the construction equipment industry, which became dominated by large firms producing more and bigger machinery than ever before. That technological innovation combined with battlefield experience to alter how builders approached earthmoving work. Before the war, builders typically bypassed steep slopes, wetlands, and other areas that required extensive hydrologic or topographical modification. But such impediments were no longer the obstacles they once were.[51]

Domestic earthmoving-equipment manufacturers such as International Harvester, Caterpillar, and LeTourneau had all held wartime government contracts worth hundreds of millions of dollars. Their plants produced battlefield icons like the M-7 tank. Other key products included heavy-duty scrapers, graders, and bulldozers that built military bases and fortifications. Thanks to war capital, Caterpillar more than doubled its factory floor space, and, like LeTourneau and Allis-Chalmers, opened foreign manufacturing centers. After the war, these corporations further expanded their domestic plants and put new technical knowledge to use in the peacetime economy. International Harvester used its experience building tanks to design powerful new tractors with treads instead of wheels. In 1947, the company started an "Industrial Products" line that included high-powered treaded bulldozers, front-end loaders, and cranes. Treads had proved vital in overseas battle conditions, more capable than inflated innertubes and rubber tires in rough landscapes and less prone to getting swamped by mud or beached on uneven ground. Nothing was more important for interstate highway construction than the continued advancement of bulldozer technology, which was the foundation of the "Construction Equipment Division" opened by International Harvester in 1956.[52]

The earthmoving business was booming. Even automobile companies got in on the action. In 1953, General Motors purchased the Euclid Road Machinery Company, which specialized in wheeled dump trucks, tractors, and scrapers. This was a dramatic example of vertical integration. Already a heavy investor in the petroleum industry, GM now had a stake in everything from the oil sales that funded the interstate system's construction to the machines

FIGURE 10. International Harvester bulldozers on an assembly line, 1959. Wisconsin Historical Society, WHI-85753.

FIGURE 11. A construction worker maneuvers a new International Harvester bulldozer model on a highway construction "field test," 1961. Wisconsin Historical Society, WHI-38794.

that made it possible to the cars rolling along its pavement.[53] New products began to roll off assembly lines, too. A significant innovation came with the Melroe Manufacturing Company's 1962 release of a compact, four-wheeled (and later treaded) skid-steer loader that it branded the "Bobcat." This small but powerful vehicle proved so successful that its design remains relatively unchanged today. The tractor-like machine featured a bucket—and later other attachments—that could be raised over the head of the operator, allowing for a reduced wheelbase and overall size. It was ideal for cramped jobs. One provocative advertisement, featuring a tightly clothed woman sucking her index finger, introduced the Bobcat as "a lovely little homewrecker." The double entendre alerted buyers to the key role the Bobcat would soon play in interstate development. In neighborhoods marked for "slum clearance," it was well suited to gutting buildings and widening alleys before larger machines moved in to clear and pave over entire neighborhoods.[54]

The connection between World War II and the interstate highway project extended to the men running the machines that invaded the American landscape. Many were veterans who had learned how to clear and grade land

FIGURE 12. A Euclid scraper removing clay to prepare for highway construction in Massachusetts, 1956. Courtesy of the Historical Construction Equipment Association Archives.

FIGURE 13. An Allis-Chalmers earthmover working on Interstate 94's construction in Wisconsin, 1959. Wisconsin Historical Society, WHI-40397.

under invasion conditions. The massive motor pool required to transport the American military and its supplies overseas meant a surplus of skilled drivers and mechanics was on hand after the war. Accustomed to following orders, they were ready to run the machinery necessary for interstate construction. They worked, as one managing engineer put it, with a "get all you can while you can" attitude—their fast, destructive approach a product of wartime incentives for clearing as much terrain as possible in the shortest amount of time.[55] The mental and bodily attributes attained through military service fit the rough work. Physical ability and a chin-up culture of obedience helped ensure expedient work under conditions that ranged from the Louisiana bayou, where temperatures along Interstate 10's path pushed into the triple digits, to Interstate 29 through the Great Plains, where subzero temperatures were common.[56]

At first, the dramatic reach of new construction and expansions of existing highways drew enthusiasm and optimism in public discourse. From the Utah desert came hopes in the *Deseret News* that Comb Ridge, the 120-mile-long, 700-foot-tall monocline running north–south in the state's southeastern corner, might soon be "conquered" by interstate highway construction. The astounding geological formation had proved "too much for

FIGURE 14. US Marines using an International Harvester bulldozer to clear ground on an active battlefield, Namur Island, 1944. Wisconsin Historical Society, WHI-24681.

the indomitable pioneers" of the nineteenth century; maybe the new wave of bulldozer-riding frontiersmen could break a path through. In Arizona, the *Phoenix Gazette* reported on a portion of interstate highway construction that traversed Virgin River Gorge. This "monumental feat" subdued the previously "impassable" area and its "sheer canyon walls soaring hundreds of feet into the air." In the Allegheny Mountains, "workmen are moving a mountain," wrote a breathless *Washington Post* reporter, who characterized the effort as "one of the nation's—and the world's—most gigantic and spectacular highway building projects."[57] Across all these environmental contexts, *Life* reported on "treedozers that pluck[ed] out good-sized trees like radishes," and "rock crushers that crush[ed] 150 tons of rock an hour." The heavy machinery blasted through mountains, leveled hills, and demolished neighborhoods "on a scale heretofore unfeasible."[58] Fitting the pioneer narrative, the BPR dramatized such activity with photographs showing interstate construction blasting through the Rocky Mountains, conquering Donner Pass, and blanketing Los Angeles as it neared the shores of the Pacific Ocean.

FIGURE 15. Interstate 70 cuts through the Rocky Mountains near Denver, 1959. Courtesy of the Federal Highway Administration.

FIGURE 16. Interstate 80 crosses Donner Pass, 1965. Courtesy of the Federal Highway Administration.

FIGURE 17. Interstate 10 and the Harbor Freeway take over Los Angeles, 1963. Courtesy of the Federal Highway Administration.

Images of dramatic engineering victories and heroic construction tales made good copy for newspapers. Inside, their business sections began to offer highway-related investing tips. The *Chicago Daily Tribune* noted an uptick in "cement stocks, earth moving machinery stocks, pneumatic tool stocks." The *Tribune* recommended the cement stocks as the "chief beneficiaries" of the interstate project, favoring well-known companies such as General Portland and Marquette. Concerning earthmoving machinery, the *Tribune* suggested taking positions in Allis-Chalmers on account of its industry edge in "motor graders, tractors, and other road building equipment." Other notable inclusions in investors' guides included concrete mixers made by Jaeger Machine and Westinghouse Air Brake, and paving equipment from Warren Brothers.[59]

Meanwhile, the very process of building the interstate highway system contributed to the Highway Trust Fund's operation. The gasoline and diesel that ran construction equipment also funded the project, and the more the machines cleared and paved, the faster new miles of highway would open to vehicles paying their way in petroleum excise taxes. The interstate highway project was the physical embodiment of a policy feedback loop. If politicians were not captured within the loop, they were at least captivated. It took little time for the full force of the interstate highway project's path dependence—its ability to chart the path of its own endurance—to be felt. In January 1958, the Department of Commerce reestimated the cost of the interstate project to be $37.6 billion, with the 90 percent federal share amounting to $33.9 billion. A year later, those figures had climbed to $41 billion and $37 billion, respectively.[60] When the president of AASHO requested that Congress deliver additional funds to kick-start construction in 1958 and compensate for the lag while excise taxes pooled up in the Highway Trust Fund, Senator Albert Gore Sr. led the way in appropriating $1.8 billion from general revenue. AASHO's president justified his request on the grounds of ensuring no delay in the construction of "a continuing, expanding future economy."[61] The next year, the Federal-Aid Highway Act of 1959 added one cent to the excise tax on gasoline, raising the total to four cents per gallon. Unlike the 1956 highway act, the 1959 increase passed through Congress easily, with President Eisenhower pushing for it.[62]

Alongside the rising costs came mounting concerns about corruption and overbilling. When President Eisenhower signed the 1959 act, he appointed retired Army Major General John Bragdon to investigate construction methods. On September 4, Speaker of the House Sam Rayburn (D-TX) appointed Rep. John A. Blatnik (D-MN) to lead a new special subcommittee on the federal-aid highway program to investigate the possibility of corruption in the interstate system. If these were well-intentioned attempts to build executive and representative oversight into interstate highway construction, they were also still further efforts to craft the public transcript, reassuring the public that there was nothing to see in the inner workings of the perpetual-motion machine.[63]

*

When the Tax Institute of America convened after the passage of the 1956 highway act, the question of cooperation in American state building had been paramount. Even as highway officials crafted a common culture and incentivized local and state governments to carry out the federal plan, subnational officials might have strained for more authority than they did. They might have dragged their feet even as they lined up, a phenomenon not uncommon in other policy areas with federal funding on the line, such as welfare.[64] That subnational highway officials did not is testament to the power of the combined structural and cultural forces animating the interstate highway project.

Pavement flowed fast thanks to the absence of formal political disagreements. That absence also helped to obscure, for a time, other clashes of more grassroots varieties. But keen observers anticipated that such clashes might emerge. Edward Lutz, the public administration professor from Cornell, noted the lurking complexity of the question "Who should decide where the highway goes?" Engineers, local officials, city planners, federal administrators, community members? Democracy, power, and domination were at stake. Lutz had little idea who would triumph in the battle to determine the paths of the nation's infrastructural future, but he was confident that this messy complex of actors would pose a distinct "hazard" to the success of the interstate highway system.[65]

Frank Moore, the Government Affairs Foundation president, noted that "many smaller communities" would be "bypassed" by the planned routes—left off the map of supposed economic integration and growth. Would they lobby to bring the interstates to their communities? Or would they be glad to be spared the potential disruption of construction? It was already evident that many communities would be obliterated in the name of progress.[66] Moore noted that the 1956 act included provisions for "public hearings," but it seemed unlikely these would give local inhabitants a voice in the development of an infrastructure system whose ambition was to remap the circuits of American life. Whether Moore knew it or not, he identified the problem that would become stark when bulldozing and paving got underway: The aggregate economic benefits that officials obsessed over would not necessarily be felt in local spaces. It was here that strife was likely to emerge, in the locales where the national benefits and local pitfalls of the interstate project became apparent in daily life.

What was clear to observers like Lutz and Moore as early as 1956 was that the interstate highway project depended on governmental function as much as on the availability of capital, asphalt, and concrete. Lutz concluded his paper with the rallying cry that the Clay Committee had offered the nation a year earlier: The United States was "a nation on wheels." And, Lutz continued, "We cannot permit these wheels to slow down." To be successful, the interstate system would need to not only "keep those wheels spinning, but add some wheels within wheels." He was referring to the practices of governance and the multiple moving parts that would need to act in a coordinated and continual fashion. Lutz ended with an invocation of Alexis de Tocqueville: "A democratic people tends toward centralization, as it were by instinct."[67] It was not clear if this was a warning or yet another assurance that Americans need not worry about statism.

Over the next decade, stretching into the 1960s, the federal government made a series of administrative changes that solidified central authority and control over the interstate highway project. The first changes came so fast they went almost unnoticed. In August 1956, just a month after the Highway Trust Fund began its appropriation and expenditure cycle, the Eisenhower administration added a key subcabinet position: federal highway administrator. The position had a symbolic valence. Many of AASHO's key state officials were governor appointed and operated with the confidence and clout born of the authority conferred on them. Eisenhower—acting on the advice of Francis du Pont—believed that a federal counterpart carrying the prestige of presidential nomination and Senate confirmation befitted the interstate project. He convinced Senator Albert Gore Sr. (D-TN) and Senator Prescott Bush (R-CT)—who had been adversaries in the interstate system's legislative design—to support the plan. The administration's choice to serve as the inaugural federal highway administrator was Bertram Tallamy, a celebrated civil engineer who had overseen the development of the New York State Thruway. Tallamy's nomination indicated that the new administrative role was functional as well as symbolic. He had managed a complicated construction process in a densely populated state, navigating issues of local and state power that would be amplified in the interstate project.[68]

The functional significance of the new role came on further display when Tallamy had to delay taking up his post (Thruway commitments kept him in New York until 1957). In the interim, Eisenhower appointed John A. Volpe. Like Tallamy, Volpe had long experience in state-level infrastructure development, having recently resigned as the commissioner of public works in Massachusetts. Like New York's, Massachusetts's intricate set of subnational governing authorities made it a useful microcosm for the kinds of issues that seemed sure to appear in the interstate construction process. Volpe had also spent three years serving with the Army Corps of Engineers, an experience that primed him to start an engineering firm that counted shopping centers, university buildings, state offices, and even military installations in its portfolio.[69]

Volpe and Tallamy carved out a role for the federal highway administrator that blended advising the executive branch on highway legislation with setting and administering the policies of the BPR. The BPR retained its commissioner of public roads position, which was now tasked with overseeing "day-to-day operations." Among Volpe's first official actions was to institute a policy of "decentralization," an ambition resonant with the Eisenhower administration's talk of smaller and more efficient government. Decentralization was a politically convenient misnomer for the changes Volpe set in motion. In this case, it meant dispatching federal BPR administrators to "field offices" from which they could ensure that localized routing and construction operations conformed to the federal plan. State and local officials would not be alone on the front lines.[70]

In a series of procedural memoranda circulated to local, state, and federal apparatuses, Volpe averred that threading this pattern of federal bureaucratic management throughout the nation was the right way to "expedite construction and completion." BPR officials arrived in the field offices with enhanced authority, empowered to "take final action" on any "matters that were in the past referred to Washington for decision or approval." The intentionally vague wording enabled significant power: the authority to grant general approval for state proposals for interstate route selection, and to select the final location of interstate segments between cities. BPR administrators also retained authority over "personnel and fiscal management" of the field offices. Volpe, navigating the ever-present tensions between national and subnational government, admitted that such "transfers" gave the BPR "virtually complete operating authority."[71]

The trend continued in 1962 when Rex Whitton, the federal highway administrator under President John F. Kennedy, made procedural changes to the role of "division engineer," the top BPR official in each state. This had typically been a stable appointment, reflecting the traditionally unfraught federal-state partnership in road work. Whitton, however, began reassigning many division engineers to new states to combat what he perceived as a tendency to "go native" and work too closely with state and local personnel. With such administrative changes, the BPR fortified central planning and federal authority.[72]

Still, the BPR assured the public that "the Federal-aid program [was] a cooperative one." The various officials—local, state, and federal—were "partners." There was a measure of truth to such claims, though they shrouded just how much power federal officials held.[73] The *Engineering News-Record*, a trade publication for highway-linked industries, reported blandly, if insightfully, that the BPR was evidently the " 'senior partner' in the federal-aid program."[74] No one seemed concerned.

A few years later, the BPR's capacity and authority expanded still further. The Department of Transportation Act of 1966, signed into law by President Lyndon B. Johnson, created the Department of Transportation (DOT) that exists today. The DOT created a coordinating structure for thirty-one transportation-related agencies, including the BPR—which it renamed the Federal Highway Administration (FHWA). The new terminology matched the federal highway administrator role created a decade prior. The FHWA was the centerpiece of the reorganization: The DOT's initial annual budget was $6.6 billion, with the FHWA accounting for $4.4 billion of this sum. Perhaps in an effort to maintain the legibility of agency activities for the public, an entity called the Bureau of Public Roads lived on. The title now referred to a subagency within the FHWA whose personnel would continue to oversee some aspects of the interstate project, including public outreach efforts. But command now resided in the FHWA, whose staff increased from 4,800 employees at the end of 1966 to 5,360 employees a year later.[75] The changes added up to heightened federal authority over transportation infrastructure and increased capacity to wield power.

*

One of the markers of the public transcript of government activity, according to James Scott's theory, is "the appearance of unanimity" crafted by elite power wielders. This veil of hegemony helps to quell popular dissent, and it also works to produce a state-making culture from which officials themselves cannot easily deviate or desist.[76] The demographic homogeneity and clubby atmosphere of institutions like AASHO, the BPR, and the FHWA served a purpose; perhaps it should be no surprise that when the race- and class-charged goal of "slum clearance" was the federally aided ambition rather than, say, poverty relief, subnational officials had little problem working with federal blueprints and oversight. In addition to fiscal incentives and the likelihood that a shared social vision contributed to the spirit of cooperation, the performance of coordination and unity became a self-fulfilling prophecy: Once articulated, official plans could not be abandoned or discredited without risking political interference of grassroots and formal varieties. The result was ideological path dependence that shaped policy adherence—and inequitable social outcomes.[77]

Officials everywhere understood what was at stake. David Stevens, chairman of the Maine State Highway Commission, reminded AASHO's members that the 1956 highway legislation, despite all its patriotic whitewashing, was a "highly controversial act" that lacked "complete unanimity of thinking." It was a telling concession that the official attitude of harmony in the interstate highway project's public transcript was a performance and not an organic arrangement. The 1956 act's consequences carried "the residue of opposition" that had been present "from the very inception of the law financing the construction of the Interstate System."[78]

Stevens understood the dramatic power now cohered in the Highway Trust Fund and in the unanimity of highway officials. But he also understood this power was not sufficient to wipe away that residue of opposition. Fiscal and institutional means would have to join with other techniques of state building to justify literally earthshaking events. Stevens foresaw problems arising in rural and urban landscapes alike. Farmers would struggle with the impermeability of the interstate highways, forced "to journey to a specific interchange" to access the highway—or to cross it to access farmland made artificially distant and inaccessible by the pavement barrier. Elsewhere, mostly in cities, "those whose property is involved in the construction" would "very materially attribute their woes to the Interstate Program." Eminent domain claims would cause "aggravation" among property owners uninterested in or unimpressed by the payoffs that supposedly resulted from highway construction. He prophesied further: Communities' frustrations might "result in public non-acceptance" of the interstate system and "snowball to the point of even jeopardizing specific projects." Public acceptance would be hard to come by—at least until construction was complete. After that, there would be little that anyone who objected could do.[79]

Stevens also recognized—and perhaps even valued—the absence of democratic participation in the guiding legislation. But this, too, gave him concerns. The 1956 act required that state highway departments "[afford] the opportunity" for public hearings pertaining to any section of interstate highway construction that crossed the boundaries of a "city, town, or village." There was no requirement that officials make use of anything the public told them at a hearing. Instead, Stevens recommended that administrators undertake "considerable educational work" in community settings. They should use public hearings not to listen to local needs but rather "to convince the citizens . . . as to the necessity for the expressway." In urban areas—"wherever the Interstate System [was] to be located in a compact section"—Stevens assured his colleagues that talk of "good planning" could make communities believe that "the only way to relieve congestion on city streets was by means of expressways located as close to the center of the city as possible." This was the official argument to be deployed: Tell communities there was a problem with traffic, convince them that bulldozing a path for the interstate system was the singular solution, and turn matters over to planning boards—who were "aware of the opportunities" enabled by "the federal government providing 90 percent of the cost" of construction, and would, presumably, get to work eagerly.[80]

Stevens's strategies were insidious. As he explained to his colleagues, the critical step in a given area was to get any portion of planned highway built, however small. Then, the coercive power of infrastructural development would begin to take over. Here, Stevens described how administrators could make highways agents of their own path dependence and infrastructural dominance. "Once a section of the Interstate Highway is completed[,] it, in turn, will sell a desire on the part of the public for additional sections,"

FIGURE 18. A bridge section under construction for Interstate 10 in Louisiana, 1967. Floating in the air, it awaited future highway construction to join it at either end. Courtesy of the Federal Highway Administration.

he explained. If community members objected to the construction of an interstate highway, this line of reasoning held, they would have an even greater objection to a partially completed and unusable section of highway rearing up in their midst. And, if called upon, politicians or judges would be likely to support the project's completion.[81]

The official attitude was one of condescension: "We knew what it was the public needed. And they didn't," recalled one highway official from Ohio.[82] The official plan would proceed, input free. This approach drove the BPR's policies and procedures. AASHO reported rising "demand for right of way personnel"—meaning surveyors, appraisers, lawyers, and negotiators—as soon as the 1956 legislation hit the books, with property takings and construction barely begun. AASHO began to formulate policy stipulating that "appraisals of partial takings should always be based on completed construction plans" that had been "developed to the stage where all basic elements of design affecting the abutting property have been determined." In other words, officials should push construction right up to the boundaries of yet-to-be-taken property before appraising the next parcel and acquiring a right-of-way. Property owners would have far less hope of blocking construction if right-of-way officials could argue in court that a given parcel of land was the only land capable of connecting highway segments already planned or constructed. AASHO also recommended that a single official carry out the duties of appraising land for right-of-way acquisition and negotiating the final payment with the landowner. Playing both roles would give the official a "decided advantage in negotiation."[83]

The rise of such practices—ranging from telling stories about congestion in inner cities to manipulating the right-of-way process and compensation for takings—explained Stevens's focus on the social infrastructure surrounding the blunt work of taking land and turning bulldozers loose. "Material shortages, lack of a sufficient amount of engineering personnel, [and] financing" would not be limiting factors, he told his AASHO colleagues. Instead, what worried him was the "real problem," "the matter of public acceptance." By *acceptance*, Stevens meant "educating the public" in a manner fitting official priorities. It was a task, he told his colleagues, they would need to "carry on for the entire period in which the Interstate System [was] being constructed."[84]

For the highway men, maintaining the public transcript would require constant work. And throughout the 1960s, critics and activists would counter with equally constant efforts to tear through the transcript and air out the "dirty linen" of government activity. They hoped to cast off the undemocratic and dominating manipulations attempted by men like Stevens, and to uncover the fiscal mechanisms and official priorities that had set the interstate highway project on its destructive and perhaps unstoppable path.

PART TWO

Confronting

4
Concrete Nightmares

In 1958, Lewis Mumford—perhaps the world's leading analyst of social geography and the built environment—published his first appraisal of the new interstate highway project. At the time, most Americans were captivated by its grandeur and convinced of the benefits its planners promised. Mumford was skeptical. Cutting against the era's assumptions about growth and development, he aimed to subvert the priorities and choices that officials worked tirelessly to naturalize. He pulled no punches: "When the American people, through their Congress, voted for a twenty-six-billion-dollar highway program," he wrote, "the most charitable thing to assume about this action is that they hadn't the faintest notion of what they were doing."[1]

Castigating the interstate system as nothing more than a dead-end manifestation of a "bulldozing habit of mind," Mumford's essay advanced a socio-environmental critique that built on his long career's work, stretching back to his first book, *The Story of Utopias*, published in 1922.[2] Since then, his analyses had focused on the human tendency toward overdevelopment, and on the capitalist agendas responsible for ecological devastation and social inequality. In urban contexts from New York City to Honolulu, he traced the twentieth-century propensity for covering land in pavement and bounding life within infrastructure systems.[3] Even in the earliest days of the interstate highway project, he observed that it came with no "hesitat[ion]to lay waste to woods, streams, parks and human neighborhoods." He foresaw a future in which one might pick any point on the map and find "a tomb of concrete roads and ramps covering the dead corpse of a city."[4]

Mumford's critique took aim at the "culture of clearance" that shaped government development projects in the postwar United States.[5] The builders of the interstate highway system continued to place a premium on scouring and flattening as much terrain as possible. AASHO and other highway interests worked hard to maintain cultural dominance, drawing on themes of masculinity, conquest, and authority. The aim was to promote the appealing notion of the freedom of the road rather than the destruction inherent in highway development. Yet public criticism grew. One reason the highway men paid such attention to their cultural maintenance work was because of the rising volume of discordant voices like Mumford's. Throughout the 1960s, the path he charted attracted fellow travelers who sought to bring the often-hidden governmental mechanics of infrastructural domination into public sight.

Eventually some of those critical voices would develop within the ranks of highway officials. By the end of the 1960s, Ben Kelley, the BPR's inaugural director of public affairs, would quit his post to become a sharp critic of the interstate highway project. Breaking away from the fraternity of highway officials, he crystallized a decade's worth of dissenting analysis in *The Pavers and the Paved: The Real Cost of America's Highway Program*, published in 1971. Kelley's efforts, like those of other critics, were public in orientation. Lucid yet grounded in technical analysis, Kelley's approach explained the legal and financial elements that drove interstate development relentlessly onward. He was concerned by the "bigness" of the interstate system. Its reach "extend[ed] not only over space, materials, dollars, and people affected by highways, but over time as well." He feared that the social, economic, and environmental consequences etched into the landscape would become permanent fixtures of American life. Whether "needed or not," a highway was destined to endure "for centuries once it is built," immune to reform or adaptation because of its brute physicality.[6]

These were the same infrastructural qualities that Kelley, in his BPR role, had once celebrated. In boosterish publications like *America's Lifelines*, the BPR's public relations team had aimed to fuel the "peculiarly American sort of pride that is stimulated by huge engineering victories." But now he was speaking the truth: The development underway was a "land-gulping process" that upended every environmental and social context it encountered. "Highways are routinely routed through houses, schools, forests, cemeteries, mountains, cities," Kelley noted. But "how often does one hear of a plan to route a park through a freeway, or run a lake through a road?"[7] The rhetorical question emphasized the stakes. Kelley joined other dissenters in reframing the interstate system as a set of priorities that ordered society according to a particular vision of aggregate economic growth, an agenda that came with steep downsides. If there was any hope of advancing a countervailing vision—one more attentive to environmental and social well-being, and one in which citizens had a voice—an important step was to make clear the heavy burdens infrastructure placed on many of the nation's communities.

Kelley's betrayal of his fellow highway men dramatized the battle lines that had been drawn throughout the 1960s. One on side was the "highway establishment," a category named by sociologist David Hapgood. This was the conglomeration of public and private power wielders who acted as boosters for the interstate highway project, supported the Highway Trust Fund, and celebrated its transformations of the national landscape. "Those on the establishment side were all male, white, middle-aged, close-cropped." And "on the other side," Hapgood wrote, were "the tweedy ladies who defend good causes, the black, the young and the old (under 40 and over 60), the shaggy, the old-money oddball, the intellectual."[8] Mumford prophesied problems that, in 1958, had yet to materialize. But as interstate construction began in earnest, its drawbacks became apparent. The bulldozing habit of mind took aim at parks, neighborhoods, and entire towns and cities. It became impossible to see the interstate highway system's construction separately from its twinned process of destruction.

At the peak of the interstate project's destructive effects, in the late 1960s, it demolished seventy thousand residential units annually, targeting poor and nonwhite communities disproportionately. Officials knew what they were doing. Destruction and displacement, as one conceded, were "particularly serious in the big city black ghettos where the supply of housing is inadequate and relocation beyond the confines of the ghetto is severely limited by racial segregation." In such areas, protests erupted: "Freeway revolts" sprang up nationwide, from San Francisco to Boston, from Baltimore to Los Angeles, and throughout an array of locales in between. "Freeway fighters," as some called the protestors, confronted and sought to limit the destructive outcomes of interstate development.[9]

While the highway establishment warded off many of these activist efforts, administrators knew they were losing control of the optimistic narrative they had spent years draping over the interstate project. Next, they worried, their administrative powers might start to slip away.[10] One official lamented at the 1965 AASHO meeting that communities increasingly "demand[ed] facts," and were no longer "reluctant to disagree" with engineers and administrators.[11] It was a sign of rising anti-interstate sentiment. Throughout the 1960s, the methods by which the interstate highways altered society began to draw scrutiny as never before. Critics aimed to connect the physical devastation and grassroots protests they saw springing up to the prescriptions of statutes, the circuits of capital expenditure, and the decisions of officials that made it possible for the interstate system to embed inequality so deeply in the national landscape.

*

Before critical commentators could bring attention to the undergirding mechanisms of power and authority that made the interstate project possible, tangible problems of destruction, construction, and resistance had to emerge and dramatize the situation. The earliest and perhaps most successful episode of interstate protest erupted in San Francisco in 1959. Beginning in 1943, some city and state officials had advanced plans for freeways connecting the Bay Bridge and the Golden Gate Bridge, saying they would revitalize so-called blight stretching west and south from the famous Ferry Building. After the passage of the Federal-Aid Highway Act of 1944 and the three-year-long route-planning process that followed, these ambitions were absorbed into the interstate highway system's blueprint. The new scheme called for Interstate 480, known as the Embarcadero Freeway, to complete Interstate 80's route over the Bay Bridge. The work pushed rapidly through Oakland and across the bridge after the passage of the Federal-Aid Highway Act of 1956. Opened to traffic in 1959, the interstate broached the eastern side of San Francisco, rearing up above Rincon Hill, the Theater District, and the Tenderloin. The *San Francisco Chronicle* called it "a crime which cannot be prettied up," a summation that reflected widespread local displeasure. The next step would be the construction of Interstate 280, an extension of I-80 that would complete plans for "ultramodern" highways circling westward from Oakland, through the Sunset District, and north across the Golden Gate Bridge.[12]

These plans were products of the state highway department, with input from local and federal highway officials, as was standard. The San Francisco Board of Supervisors, however, proved amenable to listening to complaints from citizens, and in this instance broke the facade of unanimity that generally marked federal highway development. Protestors' success in San Francisco had everything to do with the alacrity with which they acted—and also with a particular legal contingency. California state law prohibited the closure of roads within city limits unless approved by city officials. State and federal officials had no jurisdiction over the matter. Interstate construction necessitated local road closure, offering the Board of Supervisors—swayed by protestors—an unusual opportunity to block the project by refusing to shut down surface streets to make way for bulldozing and paving. City officials did not hide their motives for doing so. On January 27, 1959, the board adopted a resolution opposing construction of freeways within city limits, citing "the demolition of homes, the destruction of residential areas, the forced uprooting and relocation of individuals, families and business enterprises," all of which, they added, would remove property from city tax rolls in addition to causing untold social harm.[13]

It was an early and notable victory, and freeway revolts spread in the years following. Cultural awareness grew that the interstate project was not quite as friendly as its boosters would have Americans believe. By 1964, the folk singer

Malvina Reynolds had composed "Cement Octopus," an anti-highway ballad she performed at protests throughout California. Around the same time, Pete Seeger's rendition of Reynolds's "Little Boxes"—a pointed disparagement of the era's interstate-borne suburban sprawl—found success on radio charts. In "Cement Octopus," Reynolds cut to the core of the interstate highway project: With "red tape to eat" and "gasoline taxes to drink," the new highways "roll[ed] over everything in sight"—no regulations to check their progress, limitless capital to spur their advance. "Stand by me and protect that tree from the freeway misery," she implored. The ballad went on to ask why men could not be put to work instead building parks for children, and invited Americans to join together and "[push] that cement monster away."[14] And as the concrete leviathan gathered strength, so did protests, especially in the dense, clustered cities of the East Coast. Historians have taken note of Jane Jacobs's "bunch of mothers" in New York City, Movement Against Destruction in Baltimore, and Black United Front in Boston, among other significant flashpoints. These uprisings spanned demographic and political contexts and shared a common goal of injecting community needs into interstate highway planning and development.[15]

The results, however, were muted. Even when protestors succeeded in making themselves heard they did little to destabilize the dominance of highway builders. One interstate critic averred in 1967 that the San Francisco uprising was the "single exception" to the rule that interstate highway construction would prevail. No other community to that point had "successfully thwarted the bulldozer," at least in such a complete manner.[16] The question, for freeway fighters and critics, was how to build upon that example—a question that did not seem to have a ready answer in the 1960s.

*

In the early days of interstate construction, Lewis Mumford called for an "adequate system" of government to redress the errors of the project and avoid infrastructural pitfalls in the future.[17] He held a nuanced belief that it was not the presence of government power that was the problem, but the particular nature and mechanics of the administrative apparatuses in control, and the substantive goals their leaders sought. Less sympathetic critiques of government power emerged, too.

While officials had worked hard to navigate the interstate highway project around Cold War fears of robust central government, more mundane conservative quarrels with government spending also began to crop up. In July 1960, for example, the investigative journalist Karl Detzer published "Our Great Big Highway Bungle," in *Reader's Digest*. His account indicated that public disenchantment with the interstate system was spreading.[18] While farms, neighborhoods, and open spaces were irreparably cleaved by highway construction, it was the magnitude of public expenditure required by the interstate project that held Detzer's attention. Over 9,000 miles of the interstate system was already open to traffic, with another 4,700 miles in the works. Rights-of-way acquisitions were in progress on an additional 10,093 miles slated for construction. And mile by mile, Detzer claimed, "venality and graft" oozed from the system's legislative design and followed the highways wherever their pavement led.[19]

It was unsurprising that Detzer's particular take on the interstate system's flaws appeared in the *Digest*. In the mid-1940s, the publication had been an important intellectual hub for the postwar conservative agenda. Lifted by the era's rising tide of conservatism, the *Digest* had become America's most widely circulating magazine, reaching over eight million readers. Its editor, Max Eastman, had moved steadily to the political right over the course of his career, and he pushed readers to follow suit. Eastman was in fact responsible for one of the pivotal moments in the trajectory of American conservatism when, in 1945, he published an abridged version of Friedrich Hayek's *The Road to Serfdom* in the *Digest*. The abridgement was a coup for Hayek, who was at the time a relatively unknown figure in America.[20]

The *Digest* version of Hayek's work was a revision more than an abridgement. Eastman colluded with DeWitt Wallace, the *Digest*'s rabidly anti-communist founding editor, to map Hayek's arguments onto the ideological terrain of American libertarianism. It was just enough of a recasting to signal that the *Digest* intended to move "to the extreme Right," as an editor at *The Saturday Review of Literature* put it. The *Digest*'s emerging agenda was to caricature public goods as inefficient and public investment as fraud. Ironically, Hayek's actual analysis in *The Road to Serfdom* had identified highways—and transportation generally—as an area in which the state, rather than the market, did have a key function to perform. But in the postwar United States there was no beacon of state building larger than the interstate project. So as the *Digest* dallied with swapping out democracy and public goods for idealized free markets and pecuniary profits, it fell to journalists like Detzer to fit the neo-Hayekian narrative to conditions on the ground.[21]

While Mumford had framed his pathbreaking critique by suggesting naivete was to blame for the interstate system's failings, Detzer's essay came baited with implications of intentional deception. "The American people in 1956 were sold a beautiful 27-billion-dollar dream," he opened. The interstates were supposed to pay for themselves "painlessly out of our federal taxes on tires and our three-cents-a-gallon federal gasoline tax." But the dream proved to be "a nightmare." The price tag had climbed to $40 billion by the time he wrote, and "extravagance" and "bureaucratic stupidity" abounded, funded by the excise tax system.[22] Along these lines, the essay aimed to induce panic about the supposed moral turpitude of public officials. Whether or not there was a degree of truth to Detzer's claims, a more useful exercise was to read against the grain of his and the *Digest*'s priorities and discern deeper structural problems that observers across the political spectrum began to identify.

To start with, almost as an aside, Detzer drew attention to a basic material problem: Thanks to cheap oil supplies and the petrodollar mill that was the Highway Trust Fund, cash and pavement were in plentiful supply. No one in power had much reason to curtail construction. It was a perfect fiscal storm of overcapitalization and overdevelopment. By design, the gasoline excise tax revenue accumulating in the trust fund remained hidden and unobtrusive to consumers. In parallel, Congress had positioned the trust fund out of sight, separated from general revenue appropriation battles where it might have come under scrutiny. Those who knew what hid in plain sight understood the seriousness of the situation. Detzer solicited the commentary of Senator Harry Byrd (D-VA), a longtime critic of the cloaked and convoluted financing that fueled the interstate system. Byrd called the situation an "inexcusable mess" and worried, rather vaguely, that officials and private investors would fall prey to a "temptation to grab land, hike prices and profiteer" as the construction process wore on. (Although Byrd presumably did not take issue with the racial segregation the interstate system was in the midst of structuring.)[23] Few politicians had been as critical of the interstate system as Byrd, although he had pushed for the excise tax system himself back in the early 1950s as the lesser evil compared to bonds. Still, federal excise taxes had always been antithetical to conservative approaches to public works. And now, by the early 1960s, they appeared to be an unstoppable, cyclical force of government activity. This dual problem worried Byrd and animated the remainder of Detzer's essay.[24]

For Detzer, the core issue was oversight—or lack thereof. There was "little control written into the law" governing interstate highway financing. The federal government dispersed seemingly endless money to the states, enabling "local self-interest and greed—at the expense of the general good."[25] This analysis struck at the paradox of the interstate highway system, which had followed the long-standing New Deal framework of solidifying central planning and authority through local officials who maintained a modicum of control.[26] From the perspective of the central state apparatus in Washington, DC—which focused on measuring the rise in goods hauled and tracking the growth of the aggregate economy—the interstate highway system was working as intended, as was the mix of federal and subnational authority. Where Mumford might have asked if this arrangement was in fact best serving the health of local communities and ecologies, Detzer worried about inefficiency and waste.

Detzer was troubled by what he saw as needless costs to taxpayers. Complaints of "extravagance" preoccupied his prose—and though his essay emanated from his partisan positions, he nonetheless found striking problems. Dubious highway development was underway in urban and rural landscapes alike. Everywhere, it seemed, there was waste, destruction, displacement, or some mix of all three. Detzer found a particularly striking example in a corner of rural Nevada, where a three-mile stretch of interstate serving an average of only eighty-nine automobiles per day secured federal funds for three cloverleaf interchanges—one for each mile, and one of which accessed a single ranch. In Omaha, Detzer took note of a plan to route an interstate directly through the city center, despite the presence of an existing beltway that could have been brought up to interstate specifications at far less expense. The city center route required taking a hundred acres of land through eminent domain and destroying nine hundred homes and buildings. Detzer was not necessarily a proponent of a robust tax state, but even he noted that the subsequent "loss to the tax rolls" would be "heavy" and cause further devastation of the urban landscape.[27]

Detzer found a similar process underway in Wilmington, Delaware. The BPR blazed "a 300-foot swath right through town," following the segregation line and creating a barrier to integration in the process. As in Omaha, officials ignored a beltway option. Detzer again called attention to the tax issues that ensued: a projected annual loss of $115,000 from the homes and businesses that had stood in the interstate's path. Macon, Georgia, faced a similar plight. There, local critics pointed out that the interstates, counter to what their racist boosters claimed about "slum clearance" and "urban renewal," would be "blight-producing." Of course, power wielders in Macon dismissed residents' concerns along these lines and refused to pursue a beltway option. Detzer perceived the racist logic at play, and he underscored the entwined economic logic. Officials reasoned that "since the [federal] government is spending the money, it may as well spend some our way." Or, in the words of the Macon County Commissioner, why should the city "pass up an outlay of 50 million dollars"? Why indeed, when the money could be channeled to local contractors and used to fulfill a broader racist logic? A new interstate highway meant "profitable business for contractors, splendid real-estate deals, works for deserving labor, and quantities of cement, asphalt, gravel and steel to be bought from manufacturers and local dealers."[28]

FIGURE 19. Heavy machinery scours through Wilmington, Delaware, to make way for Interstate 95, 1966. Courtesy of the Delaware Public Archives.

Detzer saw subnational officials and private interests "rush[ing] to get at the federal moneybags" and leaving chaos in their wake.[29] Detzer's criticism of private interests here was striking, given his general faith in the market. But the availability of so much government capital could be seen as distorting market exchange and disrupting an efficient regime of supply and demand. Federal funds were flying thick and fast, and there was no checking power—whether in the form of government regulations or supply and demand relationships—capable of constraining the interstate project and making it a more functional and fiscally reasonable endeavor.

Detzer concluded that the absence of any form of balancing mechanism was not an oversight. He dug through municipal records, finding fragmentary yet telling evidence of just how freely the designers of the Highway Trust Fund intended capital to flow. In one example, Detzer revealed how the New York comptroller general had requested and received $25 million for interstate land purchases after less than a day's study of location and expected expenditure. In Georgia, he found an arithmetic error—was it an error?—that

FIGURE 20. Interstate 95's construction followed the racial boundary line through Wilmington, Delaware, in 1966, helping maintain de facto segregation. Courtesy of the Delaware Public Archives.

overstated an expenditure line by $19.7 million. The largess extended to the general industry surrounding interstate construction. Consulting engineers in Maryland alone had tallied up $24 million in fees since the interstate project's inception. "The temptation to cash in on a large scale," Detzer wrote, "is frequently great and sometimes irresistible." In rare instances Detzer found state-level officials trying to clean up the mess. An investigation by the Nevada comptroller general, for example, found "deficiencies" in twenty-nine out of forty appraisals for interstate land purchases that went unnoticed by the federal government. Many of the purchases had been overcompensated by as much as 50 percent of prevailing prices. Subnational and federal administrators seemed to have little incentive to reduce their own cost accounting or look twice at cash infusions.[30] Detzer tended to blame corruption for the problems he decried. While this characterization did not match the complexity of some critiques of highway governance, his capital-focused approach helped to reveal the extent of the problems stemming from the Highway Trust Fund.

Other critics took up the 1960s' growing interest in transcending blunt economic measurements of progress and its pitfalls, favoring more holistic valuations of political and economic agendas. The aim was to fully illuminate social arrangements rife with complex structures of inequality. Such instant touchstones as Jane Jacobs's *The Death and Life of Great American Cities* (1961), Rachel Carson's *Silent Spring* (1962), and Ralph Nader's *Unsafe at any Speed* (1965) indexed changing ideas about human and environmental relationships—and the state's role in producing and maintaining those relationships. These books came on the scene with captivating power. They challenged Americans to rethink ingrained narratives of progress attached to official manipulations of the environment, whether in terms of urban planning, natural resource management, or the cultural dominance of the automobile (which had, after all, become the key meeting point between people and the environment throughout the car-defined American landscape). Questioning official narratives was at the core of each of these landmark studies.[31]

The emergence of dissent was a new component of the interstate highway project. There had been disagreements during the legislative process leading up to the Federal-Aid Highway Act of 1956, but there had never been a substantive debate over the project's fundamental merits. The vision guiding the interstate highway system was part and parcel of a capitalist state-building system so deeply naturalized that a plan to accelerate production, distribution, and consumption prompted few critical questions. Mumford, a lonesome voice of dissent at the margins of the high modernist ideology suffusing the project, had little chance of changing the course of infrastructural development—and his critique appeared after the Highway Trust Fund began pumping out pavement. Even he had not quite foreseen the consequences before it was too late. As the pavement hardened, what interstate critics could do was deploy a strategy like that of Jacobs and explain what was afoot, even if they could not repair what had already been done. In this way, they shared an odd similarity with the highway builders: Both camps looked to the horizon, hoping for new realities. Critics focused on laying bare the priority-laden structural forces at play, revealing the official narrative of highway building to be little more than marketing.

Such criticism began to proliferate in mainstream publications. The interstates had once "served to unite America," wrote Richard Whalen, an investigative journalist for the *Saturday Evening Post*. Yet by the time he wrote this in 1968, the interstates seemed instead to "despoil our landscapes and our cities." Before long, he feared, "the whole damned country" would be "paved over." The title of his exposé posed a question that came to animate the era's concerns: "The American Highway: Do We Know Where We're Going?" The answer had never appeared more clear and less hopeful.[32]

In 1966, *America's Lifelines*—the annual, boosterish publication produced by the BPR—claimed that the interstate system was spreading "new freedom."[33] Did automobiles and highways produce greater freedom? The answer was not as obvious as it had appeared to be just a generation earlier, when the transformative invention of the Model T captivated the American imagination. Henry Ford's Tin Lizzies had offered the illusion, at least, of freedom—in the form of the automobile's dramatic revision of energy, space, and time. In less than a lifetime, the combination of automotive innovation and highway development ushered forth vast postwar suburbs lining the city-bound interstates. "The Freedom of the American Highway"—the slogan introduced by Ford in 1956 in celebration of the interstate highway project—amounted to a nation of isolated and alienated motorists stuck on long, tedious highway commutes.[34] With mobility came stasis and confinement, and not only for those grinding through daily commutes within the close confines of their cars. Suburbs shrank the world still more for those who did not work outside the home. Whalen wrote that "the suburban housewife stuck without a car is nowhere; in her more characteristic role of chauffeur, she possesses the same freedom to keep moving as a squirrel in a cage." And did the husbands of these housewives traverse a new frontier, as Ford's marketers seemed to promise? More likely they were "inching homeward behind 280 mocking horsepower" on an exhaust-choked interstate, envying their grandfathers who "drove a single horse," but might at least "breathe fresh air."[35]

*

As interstate dissent gathered force, postwar faith in the highway-borne, urban-oriented conveyance of aggregate prosperity eroded. There was the possibility of corruption that Karl Detzer had explored, but the bigger issue, in fact, had to do with an opposite problem: the federal prerogative to build efficiently. The official federal interest lay in minimizing expenses and maximizing pavement. As critics were coming to realize—and communities knew all too well—this meant seizing rights-of-way and bulldozing straight lines for "point to point" construction through poor neighborhoods.

"The whole theory of highway building has been to build the cheapest way per mile," noted a newly doubtful Department of Transportation planner in 1968, as the realities of this approach became evident. The racial implications were glaring. "Under the engineer's rule of building at the cheapest-cost-per-mile," Whalen interpolated, "the 'best' right of way often lies through inexpensive housing, which is likely to be low-rent Negro housing." Officials used economic and racial logics until it was too late: "We've discovered that knocking down neighborhoods may be the most expensive thing you can do," the planner acknowledged to Whalen. He was not speaking merely of dollars and cents. Slowly, deeper problems lodged in "official awareness." As one policy researcher put it, highway planners began to see "social costs." But policymakers and researchers—and even Whalen—still positioned engineers to take the blame: "The problem is that all decisions have been left in the hands of engineers. You can't just cost-account for the highway user; you have to cost-account for the whole city the road enters."[36]

The need to account for the "whole city" came on display in New York City, where officials had proposed routing highway construction through lower Manhattan. There, private interests made themselves visible, too, with Wall Street luminaries like David Rockefeller supporting planning efforts. It was easy for business leaders to invoke traffic and commuting to justify highway development in Manhattan. But as the sociologist David Hapgood noted in his aptly titled article "The Highwaymen," Rockefeller and his ilk had additional priorities. "People who work in the Wall Street financial district, just south of the route of the [proposed] Expressway, are less interested in construction of the Expressway than in the side effects it will bring," Hapgood wrote. "The Expressway will be the occasion for a lot of clearing." It would require "smashing the neighborhoods of lower Manhattan," displacing Jewish, Black, Puerto Rican, Eastern European, Italian, and Chinese communities particularly. "No body likes to have unfortunate neighbors next door," dog-whistled one of the Wall Streeters to Hapgood. And once the eradication was complete, the land would be available for "high-rise office buildings."[37] But the Lower Manhattan Expressway never materialized, one of the rare success stories for the era's freeway fighters. Hapgood credited Jane Jacobs and *The Death and Life of Great American Cities* with the victory. Her work "shook the city planning establishment to its concrete foundations," and upended assumptions about urban highway development.[38]

Elsewhere, however, destruction continued in the racist pattern Hapgood had noted. A. Q. Mowbray, author of *Road to Ruin: A Critical View of the Federal Highway Program* (1967), described the intentionality with which highway planners "slash[ed]" through "low-income housing areas" and nonwhite communities nationwide. It was a result, Mowbray wrote, of the "white power structure" that guided highway development. "This brutal, unjust shoving aside of black human beings to make highways for white motorists no doubt has played its part in nurturing the outrage against society that fires the black revolt." While interstate-highway boosters might pay lip service to the capital that trickled out for rights-of-way purchases, Mowbray was quick to point out the inadequacy of this arrangement. "Payment for property confiscated by the state means nothing to these residents, since most of them are renters.

They must simply move from slum to slum, with conditions worsening at every move as the supply of low-income housing diminishes."[39] That reality was becoming a central aspect of interstate dissent.

According to Richard Whalen, an "increasingly radical" faction was, by the latter half of the 1960s, beginning to attack not just the immediate problems posed by the interstate highway system but the entire regime that built it. This faction recognized the interstate system as a manifestation of the governing regime responsible for its presence. Radical freeway fighters began to frame the interstate system as a tool that officials used to etch racial discrimination and economic inequality into the landscape.[40] Searching for an illustrative example, Whalen found incontrovertible "evidence of a racial double standard" in the location of Interstate 40's route in Nashville, Tennessee. There, as elsewhere, local problems reflected a national pattern charted by federal legislation. Whalen worked at this intersection of scales, grounding his arguments in examinations of problems visible on city blocks. He explained how I-40 was "carefully fitted around such white institutions as Vanderbilt University and Peabody and Scarritt colleges," while it "divide[ed] and tend[ed] to isolate the campuses of three Negro institutions, Fisk University, Tennessee A. & I. and Meharry Medical College. Scores of dead-end streets [were] created in the ghetto. The only park used by Negroes in North Nashville [was] damaged and made less accessible. The only branch library serving the area [lay] on the other side of a single elevated walkway."[41]

Whalen described the physical construction of generational inequality in real time as he watched bulldozers destroy neighborhoods and their futures. But description was not enough to solve the issues, nor were solutions available within the city limits; I-40's effects were but the "symptom" of the "deepest crisis" of federal governance. Throughout the 1960s, in fact, the problems in Nashville were leading up to a key lawsuit. But outside the courtroom, critics like Whalen focused on exposing the racial double standard and the governmental mechanisms that created and sustained it.[42]

*

The troubling practices of government that Whalen and others started to identify could become sharper targets of analysis because communities—and the "human and social values" they contained—were being "destroyed at an intolerable rate." The facts could no longer be hidden as bulldozers and protestors filled the streets. And the official response was less than adequate.[43] In 1968, the FHWA introduced a process in which it held two public hearings on each section of interstate construction, the first on location and the second on community impact. Whalen hoped this change might strike "at the heart of the highway establishment's power," but he found it met "stiff resistance among congressional defenders of the status quo." And even apart from that resistance, the flaws were clear. The procedures continued to require officials to merely make a show of listening to citizens.[44] One journalist concluded that the hearings were "a carefully staged performance designed to show the audience why the route officially agreed upon in private cannot be changed."[45] Officials were under no obligation to act on information presented by the public.

Still, the shifting emphasis on community participation raised important questions: Did the composition of the public audience matter? If there was any chance of officials acting on what they heard, whom would they be most likely to listen to? And who, among the thousands of highway-harmed citizens, would critics take seriously, and boost onto the public stage? Whalen, for example, considered the role played by Nashville's Black leaders, whom he noted were "impeccably middle-class and moderate." Their respectability, in his telling, had helped them gain a certain level of recognition, but he also derided them for taking the "bait" from Nashville Mayor Beverly Briley, who made the standard promises of widespread economic uplift, even for the Black community. Briley had been happy to direct federal funds toward destroying poor, nonwhite neighborhoods. Protestors there began to express to Whalen that "the attitude of the white establishment" gave "the violent" among them reason "to do what they're dying to do." And "if another riot comes," the protestors said, "we're dead."[46]

Nashville, like countless other cities, had few options available to those staring down the destruction of interstate development. In a city that deployed two thousand city police, state troopers, and National Guardsmen to quell uprisings after Dr. Martin Luther King Jr.'s murder, "white bankers and the Chamber of Commerce" spoke of a private aid package for the African American business district but did not deliver funds. Public agencies, including the Small Business Administration and the Department of Housing and Urban Development, spent $25,000 in tax money on a study of North Nashville but similarly delivered no financial assistance. Some Black business owners joined together to form a stock company and attempted to purchase an interest in a nearby shopping center. Still, it was not clear if local businesses would be able to afford relocation.[47]

Whalen's mode of portraying racial discrimination—criticizing well-heeled Black leaders and radical Black activists in the same breath—was intended to appeal to a particular white 1960s readership. There were white villains in this story, to be sure, but few Black heroes. Still, Whalen pulled no punches in describing how the interstate system produced "separate and unequal populations," and he noted that "to many white men it seems the height of extremism for Negroes to demand a voice in the fate of their community." But his approach to reform was still steeped in that very whiteness, and bound up in political assumptions that contained a commitment to racial accommodation.[48] For example, when Whalen described the African American–led "Save—Don't Pave" campaign in Washington, DC, he noted the "neatly dressed Negroes" at the helm. These leaders shared a cultural orientation and class position with those on the majority Black Washington, DC, council, of which Whalen also took note. And these leaders stood in sharp contrast to a more confrontational manifestation of racial interstate politics.[49]

Whalen gasped over the involvement of "N——s, Inc.," an organization whose leadership included R. H. Booker, "a Negro in his early 20's, wearing a bushy Afro hair style, a pink shirt and sunglasses," who denounced "white men's highways through black men's bedrooms." Another group member demanded that "white men stop playing with black people's lives." This, apparently, was a step too far for Whalen. His prose grew agitated when he recounted a public hearing at which N——s, Inc. "heckled and interrupted" white highway officials, even threatening one with "a Molotov cocktail through his window." Such violence never materialized, though destruction pervaded the District in the weeks after Whalen's visit. Unrest following King's murder left parts of the city broken and ablaze, with "black columns of smoke" above the "Negro ghetto" providing a fiery reminder of the unequal urban landscape the interstate system threatened to cleave with even deeper segregation.[50]

From Whalen's vantage in 1968, it was difficult to determine a path forward. More radical approaches—those represented by N——s, Inc. and others with whom they forged solidarity—were too confrontational and perhaps too poorly organized to chip away at the conservative culture of 1960s politics. They were too radical even for Whalen, who was otherwise sympathetic to the fight. This was not a political problem limited to highway dissent. Radicalism so often failed to gain a firm foothold throughout the fractious decade. But the challenge, ultimately, had little to do with the demographics or strategies of protest. As critics took stock of harms and searched out their underlying causal forces, it was clear that the mechanisms enabling interstate highway development were well designed to repel dissent.

Freeway fighters were up against one of modern America's most powerful engines of state building: the Highway Trust Fund. Officials loved it. Harold L. Plummer, a state highway official from Wisconsin, remarked in 1956 that the fund offered the possibility of pursuing a "long-range objective" in state building for "the first time in highway history." While previous highway projects had "general objectives," their "goals were quite remote," and their funding—or lack thereof—had guided their construction at a "leisurely" pace.[51] With the Highway Trust Fund, highway administrators could build as much as they liked, wherever they liked. The steady funding made "advance planning"—which was "the most important tool available to the administrator"—possible and appealing. It was the bedrock of infrastructural state building and the antithesis of a mode of governance that might otherwise "drift" from its own ambitions and become "directionless" in its endeavors. For Plummer, this was not a statement of fact but a setting of terms for administrators. The interstate highway project's financing system allowed officials to plan long into the future, and they did their best to make their plans come true.[52]

Critics took notice. Credit went to A. Q. Mowbray for launching the first "detailed attack" on the system's fiscal mechanics, in *Road to Ruin*.[53] Mowbray, who might have been called a muckraker in a different era, argued that the Highway Trust Fund had enabled a "flow" of capital whose current was steady, predictable, and "all but immune" to the democratizing effects of legislative governance. The flow of capital provided perpetual growth measured in both revenue accrued and miles paved. The fund increased based on purchases of gasoline, oil, and tires, with revenue accumulating "independent of any new authorization by Congress."[54] This was not novel information; one could have seen as much the moment the law hit the books in 1956. But it took the better part of a decade for such statements to reach the public with the clarity Mowbray offered.

Throughout the 1960s Congress had increased the rate of excise taxation at the root of the ever-growing capital fund. Several additional Federal-Aid Highway Acts, passed in the late 1950s and 1960s, added funding, mileage, and time to the equation. The Federal-Aid Highway Act of 1968 was the pinnacle of this expansionary paradigm, nearly doubling the initial capital outlay of 1956 and dedicating an additional $21.3 billion to the interstate project.[55] The gasoline levy in the original 1956 legislation had raised taxation from two to three cents per gallon of gas, and by the mid-1960s that figure had climbed to four cents per gallon.[56] Such changes increased the total capital in the Highway Trust Fund but did not alter the accelerating rate of accumulation: This was driven by mounting consumer expenditures on gasoline, oil, and tires. From these fiscal building blocks, "this self-perpetuating fund simply grows," Mowbray concluded. As interstate mileage increased, "more cars take to the roads and drive more miles, burning more gallons of gasoline, wearing out more tires, and feeding more money into the inviolate Trust Fund. A perfect closed loop."[57]

As David Hapgood put it, there was no "scramble for federal money" like that which marked most public works projects. The earmarked excise taxes kept the "trough" running full "without legislative appropriation," and there was no political check—of either grassroots or formal varieties—on the project and its outcomes. This had long been the plan: President Eisenhower's Commission on Intergovernmental Relations reported that the fiscal machine powering highway development looked as though it would "never [reach] any definite goal"—not because it would fall short, but because it was detached from any measurable or necessary end point. The fiscal apparatus itself *was* the end point—the Highway Trust Fund was a tool of continuity with which state builders could craft political permanence. The irony was that the more physical and permanent the state-building vision became, the more legible its problems were to critics, protestors, and even some politicians.[58]

Among the many problems at hand, racism became an organizing focus of critique. Mowbray noted in 1967 that officials, when "given the choice of buildings to destroy, will naturally choose the low-income areas." Wealthier areas "hardly notice[d]" the destruction, and the land cost less to take. "Every dollar saved in right-of-way acquisition," after all, was a dollar that could be spent instead on "placing asphalt."[59] Cheaper land meant more jobs for engineers and builders, more miles of highways, and more revenue cycling back into the Highway Trust Fund. But it was not just a matter of saving money. The social priorities were inseparable and easily identified. As one Kansas highway official put it, "When you first start to look at a new location for a highway, whether it is an interstate or anything else, one of the things you always look for is low land cost." This meant that "if you are going to go through land that is developed, the underprivileged areas of town which would have a lower unit cost as far as the dwellings were concerned, always [catch] the engineer's eye ahead of trying to go through Snob Hill."[60]

Consequently, the "lava-like flow" of interstate pavement demolished urban communities everywhere; by the late 1960s, it uprooted ninety thousand people per year. And development followed "the path of least resistance," according to James D. Dilts, yet another critic. This meant "poor neighborhoods," where, he worried, "Negroes and the elderly" were particularly vulnerable. Such demographic groups, he explained, were "ill-equipped to find suitable replacement housing," the former because of "discrimination," the latter because of "fixed incomes." Whiter and wealthier groups could escape into the "formless suburban sprawl" whose growth the interstates had rapidly facilitated. But as interstate construction hollowed out urban communities, it produced "declining tax bases in already poor cities." Remaining residents would be stuck with higher costs and diminished public services compared to those who could decamp for suburbia.[61] Dilts and others were beginning to understand and explain the harms that spun out from the Highway Trust

Fund's tax-and-pave loop. Critics identified the reform-resistant character of infrastructural governance in urgent but vague terms. As Hapgood put it, "The highway revolt is against the tyranny of the machine—the highway bulldozer and the political machine that drives it."[62]

*

With their authority and good intentions in question, highway officials began to express resentment and victimhood throughout the end of the 1960s. Robert Bartlett, Pennsylvania's secretary of highways, complained that "engineering logic" was "question[ed] by dreamers" on a daily basis.[63] A state official from Minnesota made use of the 1968 AASHO policy conference to disparage "taxpayers," "interest groups," "editorial writers," and a welter of "thought leaders, citizen movers and shakers and college profs with time to reach the dreams that most of us forget when we get out of bed."[64] None of these groups, apparently, was entitled to a say in state building. Meanwhile, AASHO's president dismissed interstate protestors as a "small vociferous minority" of "so-called artistic and creative people." Such derision of interstate highway dissent was intended to maintain the status quo.[65]

But there was enough critique and animosity in the air to cause some highway officials to take notice. At the 1968 AASHO meeting, Bartlett warned his colleagues to be on guard against any political "uncertainty about the status and future" of interstate highway financing. It was not the threat of "cutbacks" in excise tax revenue that worried Bartlett—this, after all, was nearly impossible. Rather, Bartlett worried over any hint of Congress restructuring the trust fund itself. Such a scenario would require Congress to invent a new system for pooling and deploying gasoline excise taxes that were already on the books. It was unlikely to happen, especially given the added insulation of the political clash such a proposal would provoke. Bartlett's concerns were probably exaggerated. Still, his attention to the *mechanism* as much as to the capital itself was telling. A loss of funds was a problem; a loss of the ability to engage in "long-range" planning was a disaster—this was the key capacity the trust fund provided.[66]

Some degree of concern among highway officials may have been merited. There had already been attempts to alter the trust fund's operations by this time. The incident that caused officials the most worry had come in November 1966, in the relatively early days of American involvement in the Vietnam War. The White House tried to redirect some of the Highway Trust Fund's capital to the war effort but struggled to circumvent the statutory guidance that required scheduled payments for interstate highway construction. The Johnson administration claimed the plan amounted to a "freeze," and that the capital would be dispersed for the interstate system later. Of course, a freeze, whether legal or not, threatened to subvert the core premises of the trust fund and the infrastructural state building it powered. Predictable, continual funding was at the heart of the whole endeavor.

Interstate backers responded immediately, claiming the proposed freeze was not just illegal, but impractical and immoral. Ronald Reagan, the Republican governor of California, maintained that disrupting the trust fund would be "dishonest" and a "grave mistake on the part of the administration." Reagan argued that tinkering with the Highway Trust Fund would fracture public trust—in this case, a different public than that whose trust had already been shaken by the destructive process of interstate development. Rep. George Fallon (D-MD), an original architect of the Highway Trust Fund, noted that a freeze would make it "impossible to maintain an orderly highway construction program," a situation the *Engineering News-Record* warned would incur heavy "economic losses." This was a functional argument against disrupting the Highway Trust Fund. But other issues cropped up.[67]

While Democrats and Republicans had once disagreed over the creation of the tax structure at the core of the fund, defenders of it now came from both parties. Fallon was joined in assailing the Johnson administration by Rep. William C. Cramer (R-FL), ranking member of the House subcommittee on roads. Cramer accused the White House of "convert[ing] the federal-aid highway program from a stable program . . . to an 'iffy' program which goes up and down like a Yo-Yo depending upon the policies and economic theories of the administration." In Cramer's view, the very purpose of the Highway Trust Fund was to buffer infrastructural governance from political change.[68] As the matter grew serious, Fallon called for congressional hearings, at which point Johnson relented. Critics like A. Q. Mowbray were fascinated, if disheartened, by the widespread political support this episode revealed for the Highway Trust Fund. It was "untouchabl[e]" to a degree seen in few political apparatuses. After the dust settled in early 1967, Mowbray conceded "a clean victory for the forces of highway construction." The question he and others were left to ponder was, What made the trust fund so inviolable? It was not its statutory design alone. Congress could have unwound the structure, had it chosen to do so.[69]

One explanation for the Highway Trust Fund's durability is the fact that the capital it contained accounted for the vested interests of a vast array of industries and trade groups. The American Road Builders Association, the Portland Cement Association, the Asphalt Institute, the American Trucking Association, the National Association of Motor Bus Operators, and the Automobile Manufacturers Association were just the more prominent organizations tethered tightly to its capital. Ben Kelley, the former BPR director of public affairs, coauthored a blistering report on the fund with the investigative journalist Richard Hébert. They called these vested interests in the interstate highway system the nation's chief "welfare recipients." These industry groups shared capital interests and stood to benefit from interstate highway development. More miles of pavement meant more business for each. A strange alliance even existed between organized capital and organized labor. The Teamsters and the United Auto Workers stood to gain from the Highway Trust Fund's transformation of tax dollars into pavement. There was a "clear danger," wrote Kelley and Hébert, that the federal government's trust fund apparatus had tied taxation to powerful interests for whom "the perpetuation of the fund" was "an end in itself." And there was yet another large and powerful group dependent on the trust fund: federalism's conglomeration of "state highway officials, U.S. Government administrators of highway munificence and political horse traders of freeway projects in federal and state legislatures."[70]

The biting analysis that Kelley and Hébert offered came with technical appraisals of the Highway Trust Fund and broader interpretations of how its power was situated throughout society and government. They did not have to dig very deep to find evidence. The *Engineering News-Record*, a trade publication that reported on most of these industries, routinely bragged about industrial influence. At the time of the Johnson administration's failed freeze, the publication gloated that White House officials had realized "they had a tiger by the tail." They had no choice but to let it go "before the aroused beast mauled them severely . . . Contractors, equipment dealers and manufacturers, material suppliers and others lost no time in complaining to their congressional delegations in no uncertain terms." It was serious business: "The vested interests with a major stake in the program are strong, vocal, and well organized; they pack tremendous clout." While the *Engineering News-Record* spoke grandiosely of such associational power, the failure of the Johnson administration's freeze suggested this analysis was not hubris. Intervention was legally dubious and politically contentious.[71]

But even so, Robert Bartlett was sufficiently worried about the Highway Trust Fund's security in 1968 to warn his AASHO compatriots to be on their guard against attacks. His caution was still more surprising given that the Federal-Aid Highway Act of 1968, signed into law three months *before* the 1968 AASHO meeting, had added 1,500 miles to the system's plans. There were no threats on the political landscape. Still, eternal vigilance appeared to be the price of continuity. And freeway revolts, however ineffectual in local contexts, were entering public consciousness thanks to the burgeoning work of critics.

*

In his 1971 book *The Pavers and the Paved: The Real Cost of America's Highway Program*, Ben Kelley included a list of eight tactics for "how to halt a highway":[72]

> Avoid the "wait and see" approach.
> Don't lobby Congress to stop the unwanted project.
> Don't waste time developing alternatives to the highway project.
> Determine the legal powers of your local government.
> Become familiar with recent law.
> Demand a full public hearing on the proposal.
> If an initial legal challenge fails, consider appealing to a higher authority.
> Demand strict adherence to statutory guidance and agency procedure.

Kelley elaborated his advice on each point. Taken together, his eight tactics told a story of consolidated federal administrative control of the interstate highway planning and construction process. Kelley's first entreaty—to spring into action—aimed at the long-standing practice by which officials rapidly purchased whatever land they could and argued that this confined them to pursuing their favored routes. If freeway fighters beat officials to the spot and attempted to block land takings, they might have a better shot at rerouting an entire segment of interstate highway. This was among the most proactive pieces of advice Kelley offered. His subsequent strategies had more to do with identifying areas where protestors should not even bother to waste their time. The subtext was clear: Freeway fighters had an uphill battle awaiting them. But educating themselves about the state's mechanisms of infrastructural development was an important start.

Adding a measure of hope, some reforms had arrived, even if critics doubted their capacity to fundamentally alter the paradigm of community destruction and highway construction. The addition of a second public hearing was a promising change, if for no other reason than it bothered the highway men. In 1968, the American Automobile Association, worrying that more reforms might follow, took the lead in lobbying Congress not to develop statutory guidance for additional hearings or citizen appeal provisions. "It is difficult to perceive that any possible benefits" from such changes "would be commensurate with the resultant serious and costly delays in highway improvements," the association wrote. Joining the chorus, AASHO told Congress that "local lay people" should not have "a role in highway location and design." Throughout 1968, interstate development displaced approximately 55,000 people and took 130,000 acres of parkland.[73]

Among officials, Kelley was unusual in his total conversion to the side of the highway critics. But administrators could not ignore the consequences of the displacement and destruction they had wrought. Amid the countercultural tremors of the 1960s, there was "nothing like the chorus of the highway haters," said one AASHO attendee in 1966. "We must be doing something wrong."[74] A few officials were willing to at least ponder their actions. "We've been pouring concrete into a mold that may be breaking up," admitted Edward Holmes, the FHWA's policy director, in 1968. "What kinds of cities do we want? Do we *know*?" Whether he and his colleagues knew or not, some of them, at least, recognized that the permanence they had set out to inscribe came with irrevocable consequences.[75]

FHWA officials now lamented an imagined past, "when highway engineers were heroes to the American people." They commenced a novel public relations program aimed at reclaiming the promise of progress they had once embodied. But that program revealed the very problems that sustained activism and critique. Working on public outreach efforts within the FHWA, the BPR's attempts to celebrate the economic and social benefits it supposedly brought to the nation betrayed inequitable priorities. In one misguided effort, a BPR press release bragged that "the budding basketball star of tomorrow could be a kid who learned to dribble, pass and shoot because an Interstate highway came through his neighborhood." The kid shooting hoops might have been "the same youth, who wiled away hours of his life wondering what to do next." Now, he could "cavort on a basketball court laid out under a structurally modern viaduct." The racial implications were not veiled. In a scathing response, Kelley noted that the viaduct celebrated in the BPR's propaganda had "slashed through" a Black neighborhood in Wilmington, Delaware. A basketball hoop beneath an overpass was hardly compensation.[76]

Still, the BPR tried to push back against the mounting notion that "highway builders" were "people who bulldoze down every tree within sight." The agency pointed to a group of "women members" of a "garden club" in Paintsville, Kentucky, who had transplanted some saplings that were in the path of a planned interstate in Kentucky. This was proof, according to the BPR, that its environmentally destructive building practices could be balanced out with a little spadework from helpful volunteers. Officials suggested the interstates even brought ecological benefits to the landscapes they traversed. For example, the BPR claimed that on one stretch of highway in upstate New York, they had offered "thousands of ducks and birds" a "quiet nesting place" in a new median strip. The New York State Department of Conservation obliged this remarkable public relations stunt by designating the strip a "wildlife

FIGURE 21. Residents of Wilmington, Delaware, shooting hoops beneath Interstate 95, 2015. William Bretzger / USA Today Network via Imagn Images.

preserve." In South Dakota, the BPR lauded interstate highway construction for bringing "a back-to-nature bonus to a tribe of South Dakota Indians" by building "a new lake where they can enjoy fishing, swimming, and canoeing." The statement referred to a recently dammed river, choked off to make way for pavement. The BPR's press efforts, to the agency's dismay, fanned the flames of protest. One administrator from California found that when it came to public relations campaigns, "the harder we try, the less success we have."[77]

By the end of the 1960s, officials could not escape the material realities they had created, no matter how hard they tried to veil them in narratives of progress. Racism had made the problems stark. "Whether or not we believe it is justified, the black man's fury is there, and it is built on what black men see as real injustices," admitted Henrik E. Stafseth, director of the Michigan Department of Highways, as he reported on outreach efforts in 1969. "Our discussion with inner-city residents uncovered open hostility against the new highway program. They view new highways as vehicles to assist middle-class suburbanites in getting into and out of the city, with the inner-city suffering because of it. We argue that the highway program does the greatest good for the greatest number of people, but in the eyes of inner-city residents, the greatest good flows up and the bad flows to the minority."[78] Stafseth captured the issues surrounding the interstate project. The overwhelming presence of racism made a more basic problem clear: Communities had little voice in the nation's defining state-building project. Because it was such an imposing instrument of physical state building, the interstate system attuned Americans in the 1960s to the federal government's capacity to alter and shape life with little input from those whose lives would be remapped.

Contemplating these factors, some of the era's more open-minded highway officials came to see the interstate highway project as "a testing ground for our democratic form of government," as Harold King, a FHWA official, put it in 1968. It would test not only the capacity of different layers of federalism to work together but also whether the "ultimate decision by the individual"—the "engineer," "commissioner," or a "Director of Public Works"—would "meet the needs of the whole society."[79] Notably absent from this description of authority and choice-making was the participatory presence of citizens. On that front, gaining entrance to the halls of law and government—and wielding power once within—remained to be achieved.

5

No Place to Run

Flem Otey dreamed of owning a chain of supermarkets stretching across Tennessee. As a student in the late 1950s at Tennessee A&I State University, a historically Black institution in Nashville, Otey wrote a thesis on the development of the shopping centers he saw springing up alongside the new National System of Interstate and Defense Highways. He was intrigued by this new, interstate-linked geography of consumer capitalism. Highways, shopping centers, and suburbs became both symbols and mechanisms of middle-class growth that shaped the bountiful yet unequal decades of economic expansion that followed World War II. The aspirations and opportunities that the interstate highway system appeared to make possible were appealing to Otey, looking ahead to a bright future. By the mid-1960s, Otey's dream had matured: He owned a small grocery in North Nashville's Black business district and hoped to expand with a series of satellite locations along Interstate 40, which was under construction and, when completed, would connect to Knoxville in the east and Memphis in the southwest. This novel infrastructure of mobility would accomplish what highway officials had long intended, connecting major cities and spurring exchange among them. I-40 was supposed to be Otey's path forward.[1]

Otey supported the capitalist promises of the fledgling interstate system's cartography. Observing the urban protests and uprisings of the 1960s, including those in his own city, he concluded that "the entrepreneurial route" was the best way "out of the ghetto." And yet, he began to wonder if such a path was open to him. Ironically, Otey felt the interstates were instead "closing off that route." While boosters had long promised that the interstate highways would lead to economic integration and expansion, the story for Black business owners like Otey had a different ending. The problem lay with the impacts the interstates brought to poor and nonwhite urban neighborhoods. A three-mile segment of I-40 was slated for construction at Otey's business's "front door" in North Nashville. But instead of bringing customers into his neighborhood, it would overpass the area atop a massive viaduct, which would transform a hundred square blocks of homes and businesses into a vast swath of concrete pads, pylons, and foundation mounts. It would destroy "234 Negro-owned businesses" in the area, representing more than "80 percent of *all* Negro-owned businesses" in Nashville. The few that escaped physical ruin would face poor financial prospects, as I-40's construction displaced the area's inhabitants throughout the 1960s.[2] The perils looming in the neighborhood were as dramatic as they were commonplace, and dissent developed rapidly.

Communities like Otey's were the epicenters of the freeway revolts that became constitutive elements of the fractious 1960s.[3] And in North Nashville, the uprising took the form of not just grassroots protest but also litigation. When freeway fighters, there and elsewhere, turned to the law to halt construction, they encountered and challenged the mechanisms of state building that had enabled the interstate system to cause harm in the first place. The result, by the 1970s, was an instantiation of federal administrative governance that included new but narrow opportunities for citizen input and participatory democracy. Change proved to be limited and came in racist dimensions. The slightly reformed legal landscape favored environmental protection—an issue generally prioritized by whiter and wealthier freeway fighters—over neighborhood protection. By turning bulldozers and paving crews away from parkland and ever more concertedly toward neighborhoods, environmental victories were poised to come at the expense of poor communities of color. This legal trajectory took shape via the first two federal lawsuits involving interstate highway construction that reached the Supreme Court.[4]

The first case, *Nashville I-40 Steering Committee v. Ellington*, moved through the federal courts in 1967. Flournoy Coles, chair of the economics department at Fisk University—another of Nashville's historically Black institutions—worked with the NAACP to devise a legal strategy focused on civil rights and racial discrimination claims related to interstate development. Coles would soon publish a formative study of structural obstacles to "black entrepreneurship" and move to Vanderbilt University, where he became the first Black faculty member to gain tenure. The lawsuit he helped advance ended when the Supreme Court denied certiorari. A preceding Circuit Court ruling had found no evidence of discriminatory intent. The second case to reach the Supreme Court, *Citizens to Preserve Overton Park, Inc. v. Volpe*, came in 1971. Here, a local resident named Anona Stoner led the legal effort on the ground, with assistance from the Sierra Club and other environmental organizations. With carefully coiffed graying hair, a handbag, and white heels, Stoner embodied different gender and racial politics and possibilities. Though the

FIGURE 22. Dr. Flournoy Coles, right, discusses the Nashville Interstate 40 Steering Committee's efforts to protect North Nashville at a press conference at Fisk University, 1967. At center is Zephaniah Alexander Looby, a prominent Nashville civil rights attorney, professor, and councilman whose home was bombed by segregationists in 1960. At left is Harold Love, another councilman. Robert Johnson / USA Today Network via Imagn Images.

Supreme Court's ruling in *Overton Park* was not a decisive victory for environmentalism or community participation, it succeeded in blocking this instance of potential park destruction by highway builders.[5]

While *Ellington*'s outcome has been largely forgotten, the *Overton Park* decision became a canonical ruling on judicial review of administrative action. There are significant points of similarity and divergence in these cases. Neither, in the end, did much to alter governing practices. *Ellington* was a failure for plaintiffs seeking administrative and infrastructural justice. *Overton Park* made highway officials wary of groups that brought environmental claims, yet the decision nonetheless guarded the boundaries of federal administrative authority carefully. The outcomes of the two cases reinforced rather than eroded the methods of infrastructural state building in question, even as *Overton Park* helped some litigants influence policy. It is on this point that the differences between these cases offer a window into the racist landscape of state building. Read together, *Ellington* and *Overton Park* lay bare the legal and administrative practices responsible for modern American liberalism's difficulties dealing with structural issues of racial inequality. Infrastructure combined with the mid-twentieth-century legal regime to lock in racial and class hierarchies, privileging certain types of economic and political participation over others.[6]

On the ground, this meant that Flem Otey's dreams of an interstate shopping center empire dried up. By the time the investigative journalist Richard Whalen met Otey and recorded his story in 1968, Otey's grocery was one of the lucky few businesses still standing as I-40's construction cut through North Nashville. But the highway altered Otey's neighborhood irreparably. Otey estimated that he lost half of his customer base when the I-40 viaduct reared up outside his doors. Walled in by pavement and on the edge of bankruptcy, he had few options. Relocation was necessary but nearly impossible; Otey found he was "trapped by the shortage of commercially zoned property in other Negro areas and by racial discrimination in white neighborhoods." The interstates had created unprecedented mobility, yet he had "no place to run."[7]

Entrapped by America's new infrastructure of supposed interconnection, Otey identified the problem at the core of the National System of Interstate and Defense Highways—and at the core of the legal and physical state-building regime the interstate system imposed on the nation. The infrastructure project had emerged from the minds of planners who prioritized aggregate economic growth over local prosperity—an agenda inseparable from a related set of racial and class-based priorities.

*

When the interstate system existed only as a cartographic blueprint for a better future, administrators illuminated the horizons it promised with a beneficent glow. The project appeared to be post–World War II Keynesianism made concrete, carrying the promise of an endlessly expanding economic pie. But the physical construction process produced startling inequalities in local contexts. If the interstate system fixed in place the governing priorities that animated it, it also offered protestors a tangible, nation-sized object around which to organize. Those who objected to the official scheme imagined a more participatory role for themselves within the processes of modern American state building made legible by the interstate system's development.

As the pavement flowed from coast to coast, certain realities became clear. Bulldozing pathways produced physical and social destruction. The process split farms in half, drained wetlands, and blasted through mountains. In urban areas, demolished communities and displaced residents put the harmful effects of interstate construction on clear display in neighborhoods like Flem Otey's. Under cover of politically convenient slogans like "slum clearance" and "urban renewal," officials acted on their racist and classist visions, using the interstate system to displace more than one million Americans while destroying over one thousand neighborhoods in their entirety.[8]

It was in this context that freeway revolts took hold. They sprang from neighborhoods like Otey's, where demolished buildings and displaced residents made the paradoxes of infrastructural progress legible. It was against this backdrop of exploded hopes that communities began to mount legal challenges alongside street protests, aiming to recenter the well-being of neighborhoods and the voices of citizens. Rising calls for community participation followed deep intellectual and political currents that pushed for the democratization of the federal bureaucracy throughout the 1960s. The trajectory of the interstate highway system's development and reception, in fact, indexed changing sentiments toward the administrative state. Administrators had drawn up the first interstate highway plans in the 1930s, as they embodied an iteration of administrative governance that enjoyed broad public and political support. During the Great Depression, administrators and their agencies provided jobs, housing, and all manner of public works projects—the infrastructure of social stability and prosperity—at scales ranging from the local to the national.[9]

In that setting, with administrators on the front lines of unprecedented economic-crisis management, an "era of deference" had developed, in the words of legal scholar Reuel Schiller. Expertise-based regulation and administration "dislodge[d] the judiciary from its role protecting property and economic liberty from administrative agencies." By the end of the 1930s, courts were "subservient" to the administrative state now tasked with ordering and managing the nation. In this version of state action, democratic processes could identify general problems, and then administrators, with creativity and specialized knowledge, would solve them. Courts were on the other side of the equation, "inexpert" and "inflexible." It was in this context that the BPR had produced its 1939 report *Toll Roads and Free Roads*—the conceptual origin point of the interstate highway system. Federal administrators had imagined the interstate highway system at a time when the courts deferred to their plans and expertise.[10]

But that era of judicial deference to administrative expertise, authority, and action was short-lived. By 1944, just as the interstate system came into statutory existence, with the passage of the Federal-Aid Highway Act of 1944, the tide began to turn against administrative authority. That year, the Supreme Court took up the case of *Addison v. Holly Hill Fruit Farms*, which arose from a dispute in which a worker claimed a federal administrator had no authority over minimum-wage and maximum-hour exemptions for certain agricultural laborers. The broader issue at stake was the limit of administrative autonomy when it came to interpreting and implementing statutes. The court found the administrative body in question had exceeded its authority, marking a shift away from judicial deference to administrators. Then came the Administrative Procedure Act of 1946 (APA). This so-called "constitution" of American administrative law governed how agencies could issue regulations and make decisions. Crucially, the APA directed courts to review agency actions that were "arbitrary, capricious, [or] an abuse of discretion," as well as actions "unsupported by substantial evidence." The APA did not define these terms, and the manner and extent of judicial supervision of administrative action was rather ambiguous. However, the APA created a presumption of judicial review of administrative action that indicated a clear divergence from the era of deference.[11]

*

If *Addison* and the APA were efforts to check the authority of administrators, these changes to the administrative law framework were indications of broader concerns about the power of the administrative state. By the mid-1960s, even loyal former New Dealers—who had believed that judicial deference was crucial to the empowerment of a strong administrative state capable of balancing society's needs—began to question their assumptions. Take, for example, Louis Jaffe, a former lawyer for such formative New Deal institutions as the Agricultural Adjustment Administration and the National Labor Relations Board, and later a professor of law at Harvard. In his renowned 1965 book *Judicial Control of Administrative Action*, he wrote about "the dangerous tendencies of the administrative state," warning that faith in "experts" and "rationalism" could lead down a totalitarian path. "Even if we grant . . . the magnificent accomplishments of the New Deal," he continued, "we cannot forget that our age has produced elsewhere, and even on occasion in our own country, the most monstrous expressions of administrative power," referring to Nazi Germany as well as to Cold War conceptions of global socialism and communism.[12]

Some of Jaffe's fears were overblown; the United States was not becoming a communist utopia. But in his more meticulous critiques, he was not alone. In another widely read example, sociologist Philip Selznick turned his attention to one of the era's signal examples of administrative state building: the

Tennessee Valley Authority (TVA), often held up as one of the great success stories of the New Deal administrative state and headed by the liberal lion David Lilienthal. Selznick's research on the subject first appeared in 1949, at the leading edge of critical studies of administrative authority and participatory democracy. He revised his study, *The TVA and the Grassroots*, to great acclaim in 1966. Selznick acknowledged that the TVA's original blueprint included ambitions of participatory government. Yet his analysis showed how administrative units like the TVA—whether controlled locally or not—had a tendency to gather their own momentum and power and insulate their activities from democratic inputs.[13] Less than a decade after Selznick's influential revision, Robert Caro published *The Power Broker*. Caro's main purpose was to detail Robert Moses's decades-long accretion of administrative power in the metropolitan New York area. But Caro's work was also a critique of a form of administrative governance in which unelected figures like Moses could build institutions that shaped where and how millions of people lived.[14]

Critical assessments of institutions like the TVA and the many municipal authorities Moses controlled throughout his career widened critics' field of perception as they surveyed the interstate highway project. Exploring administrative overreach and citizen participation was the order of the day. Conservatives had long complained of government overreach; now, for different reasons, some liberals began to worry over the limits of democratic participation and feared that administrative governance could become unrepresentative at best and totalitarian at worst. For critics who still believed in the promise of government activity and the benefits of public goods, the basic idea was simple: Government had to become truly participatory via a social and political push that reshaped administrative law. Expert administrators needed more oversight, not less. With unstable legal and political terrain undergirding the administrative state and freeway fighters pressing for change, the interstate highway project might come under real challenge.[15]

By making the interstate highway system a technology of federal governance, the Federal-Aid Highway Act of 1944 had created the conditions for disagreement among local, state, and federal authorities, and between citizens and administrators. But it was not until the Federal-Aid Highway Act of 1956 enabled construction to begin that these clashes materialized. Basic quantification of litigation reveals the changing context. Between 1944 and 1956—a period when the interstate system amounted to little more than a map and a plan—six state-level court cases developed involving the provisions of the 1944 act. Everything changed when the Federal-Aid Highway Act of 1956 delivered funding capable of turning plans into pavement. In the first decade of construction, at least fifty-six cases developed in thirty-one state courts, while the federal appellate courts heard seven others.[16] The state cases emerged first and surfaced fundamental questions: Who should decide where to construct an interstate highway, what priorities should be paramount, and who should govern the decision-making process? For the courts, answering such questions involved settling clashes between local and national authority and evaluating the balance of power between expert administrators and affected citizens.

The statutes enacted in 1944 and 1956 placed enormous discretionary power over local route selection and construction procedures in the hands of federal administrators. This power included authorization to acquire rights-of-way on behalf of individual states if they were "unable to acquire necessary lands" for interstate highway development and authorized federal agencies to use eminent domain to seize property, as well.[17] These were the core issues that began to fill state and federal courtrooms, providing a small but clear window into the issues at hand. The problems crystallized, for example, in a pair of cases that developed in Illinois, both of which illustrated how interstate construction altered communities' place and power within federalism. The two cases brought to the fore questions about the extent of federal administrative authority—indicating the foundational questions that would soon reach the Supreme Court in *Ellington* and *Overton Park*.

The matter began in 1962, when the Illinois Department of Public Works & Buildings mapped a route for Interstate 74 that cut through a public park in Peoria. The department filed suit in the county court for condemnation of the park land. But the court dismissed the suit, citing a case decided earlier that year, *Department of Public Works & Buildings v. Ells*. In *Ells*, the department had sought to condemn school district property for highway development, but the Supreme Court of Illinois found no authorization for "condemnation of property already devoted to public use," because the federal government, in the court's view, had not directly sought condemnation but rather had done so via a state agency. The school kept its land. Surely the public park in Peoria, also municipal property, also in "public use," should receive the same treatment. But it did not. In *United States v. Certain Parcels of Land* (1962), the US District Court for the Southern District of Illinois disagreed with the Illinois Supreme Court, finding that "the interstate system . . . transcend[ed] state and local interest in the interest of a larger national purpose," ruling that federal eminent domain authorization came into effect as soon as a state decided to participate in the interstate highway project (leaving us to wonder how feasible it would have been for a state to resist federal fiscal coercion to participate).[18]

The decision made clear that national interests and federal administrative authority were paramount. And the case's outcome left state agencies to do the hard work of condemning property on behalf of federal agencies in the name of federal construction and national interest. But even as the Illinois cases indicated the supremacy of federal administrative authority that would continue to shape interstate development, the national landscape of critique, protest, and litigation attuned legislators and agency leaders to mounting frustrations and challenges.

*

By the 1960s, interstate dissenters were exploring both judicial and legislative paths of reform as they challenged administrative authority and infrastructural state building. While the interstate highway project continued to prompt considerations of the roles of citizens and judges in administrative governance, members of congress and administrators provided measured changes that tempered the freeway revolts and enabled the interstate highway project to proceed. When it came to statutory reform, Congress proved particularly attentive to the destruction of natural areas and parkland. Senators such as Clifford Case of New Jersey, Wayne Morse of Oregon, Ralph Yarborough of Texas, and Joseph Clark of Pennsylvania spoke about the dangers of unchecked highway development and lined up their votes accordingly. Clark told his colleagues that it was "time that Congress took a look at the highway program" and stopped it from destroying "spots of historic interest and great beauty by the building of eight-lane highways through the middle of our cities."[19]

Yarborough was a strong advocate. His amendment of the Federal-Aid Highway Act of 1966 required "maximum effort to preserve Federal, State, and local government parklands and historic sites and the beauty and historic value of such lands and sites." This provision barred highway administrators from authorizing interstate construction in any parkland "unless such program includes all possible planning, including consideration of alternatives to the use of such land, to minimize any harm to such park or site resulting from such use."[20] Section 4(f) of the Department of Transportation Act of 1966 reiterated the same protections and applied them to transportation-related projects of all kinds, not just highways. Two years later, the Federal-Aid Highway Act of 1968 extended the prohibitions still further, barring interstate construction in parkland "unless (1) there is no feasible and prudent alternative to the use of such land, and (2) such program includes all possible planning to minimize harm to such park, recreational area, wildlife and waterfowl refuge, or historic site resulting from such use." The parallel provisions in these three acts became known as the "Parklands Statutes."[21]

These statutory changes developed as freeway fighters prodded legislators and administrators alike. Some, such as Secretary of Transportation Alan

Boyd, reflected newfound awareness of environmental and social problems. Boyd conceded in 1967 that the "freeway revolts around the country" were "a good thing." The protests, he explained, had made clear that highway officials needed to find a "better way" to construct the interstate system than simply "tak[ing] the property of poor people" and bulldozing the "straightest possible line." Appointed by President Lyndon B. Johnson at the height of the Great Society push in 1966, Boyd had an unorthodox background and the bearing of a reformer. While federal highway and transportation officials were usually engineers, Boyd was a lawyer by training. His professional practice had focused on administrative law, which made him a valuable cabinet member as the Johnson administration surveyed the freeway revolts and the tensions they put on display between administrative decision-making and citizen participation in the communities in which protests erupted. Boyd had an important ally in Lowell K. Bridwell, whom he appointed as federal highway administrator in 1967. Bridwell also had an unusual background: He was an investigative journalist who had spent the early 1960s covering federal transportation politics. His reporting gained the attention of President John F. Kennedy's administration and then Johnson's, and he advanced through transportation-related roles within the Department of Commerce, where he became close to Boyd.[22]

Boyd and Bridwell, however, had short tenures in the Department of Transportation. With Richard Nixon's election to the presidency in 1968 came administrative changes. Nixon appointed John Volpe, a former federal highway administrator and governor of Massachusetts, to replace Boyd as secretary of transportation. Francis C. Turner, a more traditional highway engineer-administrator, replaced Bridwell. Volpe and Turner were "hard-line road builders," according to journalists, and it looked as though they were poised to "pave the country" and drop a "concrete curtain" across the national landscape. Famed economist and public intellectual John Kenneth Galbraith feared that Volpe would "cover the country with concrete." Lyn Shepard, a longtime highway reporter for the *Christian Science Monitor*, warned that Volpe and Turner planned to "stamp out" the freeway revolts.[23] Turner wasted no time defending the interstate highway project's track record. The archetypical highway builder was not a "bull-dozing maniac in a black hat tearing everything apart just for the sport of it—or out of sadism or just plain cussedness," Turner asserted. Highway professionals, not the "new breed of amateur instant experts"—lawyers like Boyd, journalists like Bridwell—knew best how to build.[24]

While Volpe and Turner sang a different tune from their predecessors, Boyd's and Bridwell's records suggested that their reformist bearing was, at least in part, grounded in publicity rather than policy. In one telling episode, Boyd, Bridwell, and Volpe testified together before Congress in 1968 regarding the increasingly litigious resistance to the interstate project.[25] One demand that had been raised was to include a more diverse range of technical experts in interstate development. For Volpe, even enriching the administrative process with other kinds of knowledge besides highway engineering was too radical a change. Volpe noted that some state highway departments were already "incorporat[ing] new professional disciplines into their engineering activities," including the contributions of sociologists, civil rights specialists, archeologists, hydrologists, and biologists. This idea to diversify expertise was "not a good one." Such professionals, Volpe averred, "really ha[d] no basic knowledge about highway design or construction. Their only basis for criticism is opinion or artistic sensibility." The inclusion of such professionals appeared to Volpe to "work against" the "very operations" of interstate development by incorporating official perspectives on issues such as community destruction and environmental disturbance.[26]

If the diversification of bureaucracy was not the right solution to the problems at hand, asked the Senate, what was? From the perspective of activists and litigants, the problem lay with decisions made by administrators. How could the process of decision-making be changed? On this point, Bridwell noted that the federal-aid highway acts gave "considerable administrative discretion to those responsible for the Federal aid highway program." And the BPR historically had placed a "tremendous emphasis" on developing "planning competence" and "removing it from the vagaries of the political system"—removing it, that is, from opportunities for citizen input or even the oversight of Congress. In Bridwell's opinion, the nation was "in a period where at least many of the public are saying, 'No, we don't want to rely upon professional planners,' be they highway, or urban renewal, or any other kind of professional planners. 'We want these decisions turned back over to the people.'" But should the agency bend to this public onslaught? Bridwell thought not: Highway planners should continue to "make the decisions," and their activities should remain "removed" from any political "influence." It seemed even reform-curious administrators were not very amenable to citizen input when put to the test.[27]

*

Throughout the 1960s, localized protests, state court cases, statutory changes, and reformist ideas had worked their way into the layers of administrative governance guiding interstate highway development. But there had not yet been a clarifying federal lawsuit that cohered the issues and asked for a solution. That began to change in 1967, when the first interstate highway case to reach the Supreme Court originated in Flem Otey's North Nashville community. There, local leaders turned to the law to claim that Interstate 40's construction was producing racial discrimination and a deprivation of due process. The case, *Nashville I-40 Steering Committee v. Ellington*, came to a conclusion when the Supreme Court declined to hear it on appeal after the United States Court of Appeals for the Sixth Circuit ruled in favor of the administrators who were responsible for selecting I-40's North Nashville route.

Meanwhile, 200 miles to the southwest, a group of mostly white activists in Memphis grew concerned about the fate of Overton Park, their city's primary urban green space. Like Otey's neighborhood, Overton Park stood in the middle of I-40's planned path. A group of residents calling themselves "Citizens to Preserve Overton Park" brought a series of legal challenges and sought instead to locate I-40's path through a nearby neighborhood where, presumably, construction would have consequences much like those Otey had experienced in Nashville. In 1971, the Memphis park protectors produced the second case to reach the Supreme Court, *Citizens to Preserve Overton Park, Inc. v. Volpe*. The contrasts between these two cases reveal the uneven priorities of state builders, the challenges of accessing administrative and infrastructural justice, and the legal and physical inequalities of state building in modern America.[28]

In North Nashville, the litigation arose from a proposed 3.6-mile stretch of I-40's route through the city. In *Nashville I-40 Steering Committee v. Ellington*, neighborhood residents claimed infringements of their Fifth and Fourteenth Amendment rights to equal protection and due process. The defendants named were Tennessee Governor Buford Ellington, Tennessee Commissioner of Highways Charles Speight, and Nashville Mayor Beverly Briley. Seeking an injunction, the plaintiffs attempting to protect their community accused these officials of "erecting a physical barrier between [the] predominantly Negro area and other parts of Nashville." They identified two central issues. First, that the administrators in charge had "failed to hold a public hearing with proper notice and failed to consider the economic effects of the proposed route as required by Section 116(c) of the Federal-Aid Highway Act of 1956." And second, that officials had selected the route either "arbitrarily or with the purpose of discriminating against the Negro or low socio-economic segments of Nashville's population."[29]

After the United States District Court for the Middle District of Tennessee heard the case and refused to grant an injunction and halt construction, the plaintiffs appealed. The United States Court of Appeals for the Sixth Circuit confirmed what was evident to the plaintiffs: I-40's construction would cause "heavy damage" to the Black community. But the judges found they had "no choice except to affirm the judgement of the District Court in refusing to grant a preliminary injunction." On the question of discrimination, the Sixth Circuit relied on the testimony of Tennessee Highway Department officials from a hearing back on May 15, 1957. An attorney for the Highway Department had claimed that with regard to "economic effects," the proposed route was "properly located and should be constructed as located." The District Court had done its duty, relying "upon the presumption of regularity of public records and compliance by public officials with duties imposed upon them by statute." Like the District Court, the Sixth Circuit took highway officials at their word and ruled in their favor. Construction could proceed, speeding the flow of traffic from one distant city to another, even as that flow ruptured the local economy of North Nashville.[30]

The Sixth Circuit also found "no proof of racial discrimination" in the selection of I-40's route, and only the hypothesis that discrimination might follow the highway's construction. Without a smoking gun, there was no evidence to support claims of discrimination and deprivation of due process and equal protection. The Sixth Circuit would avoid undermining the expertise of administrators who made substantive decisions. "The routing of highways," the judges wrote, was "the prerogative of the executive department of government, not the judiciary." Routing highways, in other words, was a substantive matter beyond the purview of judges. Surely any route selection for I-40 would bring "hardships" upon some people. But it was for "engineers, not judges" to decide where to build, so long as they made that determination within the bounds of legislative guidance and settled administrative law. Although the era of deference was slipping away, this was the role the judges would play here: ensuring that administrators followed procedures prescribing how they were to make their substantive choices—and in *Ellington*, the judges did even this as narrowly as possible.[31]

The issue of public participation, however, posed problems that did concern the Sixth Circuit judges. They were satisfied that administrators had followed procedure, but they were willing to state their opinion that the procedures were "unsatisfactory." The Federal-Aid Highway Act required administrators to hold public hearings and listen to citizens' input before deciding where and how to build. This offered a pretense, at least, of citizen participation. Yet in Nashville, administrators had made a series of suspicious choices regarding how they publicized those hearings. They posted notices only "in a 'white' neighborhood near the predominantly Negro community" that they sought to bulldoze for construction. The notices also listed an incorrect date for the hearing. When the hearing did occur, officials did not transcribe questions from the audience—only their own statements and answers.[32]

Despite all this, the judges found that "neither the statute nor the regulations of the Bureau of Public Roads prescribed how notice of hearings should be given." One might think that an erroneous date would be grounds for review and intervention. But here the judges turned, ironically, to a substantive point: Despite the error, the hearing had been "well attended," though they did not say by whom. In their deference to administrative authority over the entire process, they went so far as to say that "no literate citizen of Nashville" could have been unaware of the hearing. It was with these habits of mind and legal choices that the judges of the Sixth Circuit did nothing more than raise an eyebrow at how highway administrators had selected I-40's destructive path through North Nashville.[33]

The plaintiffs in *Ellington* continued their fight, but the Supreme Court declined to hear their appeal. The choice to deny certiorari was an interesting one. The *Addison* decision—now on the books for over twenty years—offered a roadmap for evaluating administrative discretion and overreach. With construction complete in some parts of the city and about to commence in North Nashville, the justices may have been wary of hearing the case at such a late stage. More than a decade had passed since the suspicious 1957 hearing. The physical infrastructure looked, by this point, like the foregone conclusion highway officials hoped that it was.[34] Whatever the court's reason for turning down the appeal, letting the Sixth Circuit's ruling stand appeared to be a victory for highway administrators and the administrative state more generally. If their procedures gave citizens little opportunity to participate in administrative processes that would dramatically reshape their lives, that was not a matter for courts to evaluate. Such alterations of society, in this judicial perspective, were the purview of the political process of legislative change.

Yet rather than emboldening state and federal administrators, the outcome in *Ellington* set them on edge and led them to consider more inclusive approaches to highway development. Internally, in their briefs, memos, and annual meetings, interstate highway officials were worried: They felt that they had verged too near the precipice of their power. In one notable subsequent example, the Michigan State Highway Department spent 1968 conducting "research into the growing revolt against construction of urban highways" and issued a report summarizing its findings. The officials noted that protestors had continued to succeed in halting "one highway after another" with grassroots tactics and litigation that slowed construction, even if it failed to stop it entirely. They found that people were "sick and tired of having establishment programs imposed on them." The report concluded that many communities enjoyed new highways but took issue with the "devious means" by which the infrastructure arrived. This might have been wishful thinking among administrators searching for silver linings suggestive of basic support for the interstate highway project. Whether through self-interest or a genuine spirit of reform, they sought to "develop a positive remedial program" of administration capable of keeping the interstate project on track. Still, their final pronouncement was striking. They found that community residents wanted "a piece of the action and a voice in their future."[35]

Henrik Stafseth, director of the Michigan State Highway Department, presented the report's findings at the AASHO conference in 1969. Stafseth outlined the broader governmental issues involved. He pointed out that the interstate highway project was the "most visible" aspect of the administrative state. And public trust in this highly legible aspect of administrative governance had eroded, for a host of reasons identified by Michigan survey respondents. They reported construction vehicles running rampant in small neighborhood streets, "endangering children, cracking foundations and kicking up storms of dust." Unregulated demolition crews blasted and bulldozed through the night so that paving crews could follow in the morning. They burned their rubble and trash wherever it accumulated, filling the air with smoke and pollutants. Some residents complained of "undesirables" squatting in abandoned homes as the disruptive construction process encroached. The community members who reported these issues felt "put upon by the establishment and view[ed] the highway program as just another means of maintaining dominance over people."[36]

What was the solution? Stafseth warned against the common administrative argument that improved designs and construction practices were all that was needed to temper community discontent. Such an approach, he counseled, was little more than "another imposition of establishment ideas." A comprehensive solution demanded finding ways to give "people an opportunity to participate." They needed "a voice" in the state building that surrounded them. Otherwise, they would continue to fight against "big, impersonal agencies" that "fail[ed] to recognize and respond to their personal needs." He anticipated that "a real breakthrough" in administrative state-building practices would come only if administrators "learn[ed] to relate positively to the needs of the people."[37]

The Michigan study suggested that, even if only to avoid delays brought about by lawsuits, some administrators were ready to rethink urban issues and public participation. They had good reason to do so: The relationship between the administrative state and the courts was growing unsettled. Although *Ellington* suggested the continued primacy of administrative expertise, surrounding legal shifts threatened to open new possibilities for litigation. Mindful of this, agency lawyers began to consider strategic responses to a changing legal context that appeared primed to widen standing for lawsuits directed at administrative actions.

Edwin Reis, assistant chief counsel for the FHWA, compiled a report centered on questions about whether or not courts could intervene in existing practices of agency action. Regardless of outcome, *Ellington* had indicated that highway development was on a collision course with this legal issue. Reis made it plain that while administrators may have emerged victorious in *Ellington*, the war was not over. He called attention to another 1967 case, *Abbott Laboratories v. Gardner*.[38] The ruling in this case was important because it widened standing, broadening who could sue federal agencies and under what circumstances. Reis explained *Abbott Laboratories* thus: "If someone is complaining that Federal action injures him, a Federal court can consider the case unless it is very clear that the Congress intended that there be no judicial review. The mere absence of review procedures in the statutes does not show that there was to be no judicial review."[39]

This was a significant interpretive step away from the legal tendencies of the previous three decades, in which judges had repeatedly deferred to administrative expertise.[40] *Abbott Laboratories* showed that judges were willing to take on a larger role for themselves following the general judicial review provisions included in the foundational Administrative Procedure Act of 1946.[41] And the door was open for citizens to continue pressing claims as they had in *Ellington*. Reis concluded that it was "plain that Federal courts will continue to feel that they can review the regularity of Federal administrative action." The need to remind administrators of the point underscored the significance of the change.[42]

If standing and review were revised, the civil rights and discrimination issues that had been at the center of *Ellington* might take on new life with a judiciary less deferential to administrative expertise and more willing to think about substantive issues. Reis warned highway administrators that "the area of civil rights" law was likely to reshape court cases involving "government-aided construction." Housing claims were poised to draw the attention of judges, should interstate development "force a minority group to seek housing in a segregated housing market or force a minority group to seek housing outside of a city." Courts, he thought, would "accept jurisdiction to hear such a case." It was a striking point that, among other things, acknowledged the prevalence of de facto segregation. Citing *Burton v. Wilmington Park Authority* (1960) and *Jones v. Alfred H. Mayer Co.* (1968), Reis determined that it was "clear that operations on any Government financed or aided project can be enjoined where such operations perpetuate discrimination or otherwise deny equal protection of the law." Civil rights, discrimination, and other "areas where laws have been passed to protect a particular group" were not the only issues that worried Reis. *Abbott Laboratories* helped produce a legal landscape in which "anyone who will be affected by a project" had "standing to go to court and challenge federal grants or approval of a project."[43]

Standing got plaintiffs in the courthouse door. The outcome of litigation was another matter. Still, signs pointed toward an elevated role for citizens in challenging the expertise and choices of administrators. If there was a judicial inclination to hear these challenges, agency lawyers like Reis concluded, there was a matching inclination to take the substance of such challenges seriously and work the substantive issues into the existing framework for reviewing procedural choices. Reis was optimistic that courts would continue to defer to agencies on substantive matters. But judges were beginning to take a harder look at administrative processes. They were threading the needle, weighing administrative expertise and authority on the one hand, calls for greater citizen input on the other. Without wading into the dangerous waters of substantive decision-making, judges could widen opportunities for plaintiffs to make their claims and demands. Reis reported that most plaintiffs in recent federal cases involving administrative law had attempted to "show that procedural due process was not followed."[44] The strategy hinged on arguing that administrators had not made clear how and why they took actions—opening space, however narrow and procedural, to weigh the substantive impact of those actions.

Reis offered a warning. If citizens' "right to know" how and why actions were taken was included in "procedural due process," administrators would have to take utmost care in publicizing the "procedures and standards" of their action. If they did not, substantive issues might slip into arguments over procedures. This was what had nearly derailed administrators in *Ellington*, when the courts realized how unsatisfactory hearing publication procedures had been. Another such episode might not be ruled upon with such favor to administrators.[45]

Reis closed by noting how "pessimistic" he was. "Those seeking to challenge the actions of the agencies we work for will not be stopped at the courthouse door," he concluded. The era of deference was over, having given way to what he named "the age of protests." The writing was on the wall: The "actions of the Federal Highway Administration" would be challenged "increasingly" through litigation.[46] He predicted, with keen accuracy, that such litigation would find the strongest support "in the conservation area." This emerging body of law had just received an enormous boost with the National Environmental Policy Act of 1969 (NEPA). NEPA established a sweeping set of statutory guidelines that strengthened the already significant provisions for parkland protection in interstate highway legislation. Reis asserted that "any group claiming an interest in preserving the ecology, recreational value or beauty of an area" would be "granted standing to contest whether these factors were properly considered in approving a project." If the discrimination issues at the heart of *Ellington* had failed to give citizens voice and expand judicial review, environmental concerns would succeed in doing so.[47]

A year later, in 1970, Reis's concerns about expanded review of administrative actions involving environmental change were at the center of his next report to highway administrators on the interaction of the courts and the administrative state. He wrote that "recent developments in the law only mean that the administrator may, and indeed must, realistically consider the total highway effects of his decisions on the environment and not just the effect of these decisions upon fast, safe and efficient transportation." This lawyerly rhetoric veiled a momentous shift. The need to "only" consider "the total highway effects" on the environment was no small matter. There were any number of issues within this framework that community members might now use to bring suit. There was a very real possibility that the environmental movement was going to produce a "vitiation of administrative choice" with the help of judges who sought to widen standing and the scope of judicial review of administrative procedures. The conclusion Reis drew was that "administrative action must meet the *substance* and procedural requirements of law and regulation."[48] To counter this, the safeguard Reis identified was simply for administrators to get ahead of any substantive issues. The example he pointed to was a federal case developing in Tennessee: *Citizens to Preserve Overton Park, Inc. v. Volpe*. It included the lone injunction issued up to that point in an interstate highway lawsuit; soon the matter would become the second interstate highway lawsuit to reach the Supreme Court.

*

The trouble began in Memphis when a group of citizens took issue with I-40's planned route and questioned the authority of highway administrators to carry out construction. Where discrimination claims had not proved sufficient to prompt judicial intervention in *Ellington*, the freeway fighters in Memphis had an environmental claim that held more promise. The Parklands Statutes prevented the US secretary of transportation from authorizing the dispersal of Highway Trust Fund capital for the construction of interstate highways through public parks if a "feasible and prudent" alternative route existed. If no such alternative existed, construction in parkland could occur only with "all possible planning to minimize harm." Vague though the language was, the plaintiffs grabbed on as best they could. Their aim was to devise a strategy that would stop Secretary of Transportation John Volpe from authorizing capital expenditure and construction of a six-lane interstate highway through Memphis's Overton Park.[49]

Overton Park, a 342-acre area near the center of Memphis, contained recreational attractions including a zoo, a golf course, an outdoor theater, an art academy, and nature trails through 170 acres of forest. Residents joined with the Sierra Club to claim that the planned route for I-40 would destroy 26 acres of the park. Their claim that this violated the Parklands Statutes was rejected first by the United States District Court for the Western District of Tennessee, and then by the United States Court of Appeals for the Sixth Circuit. I-40's construction had reached the borders of Overton Park in 1966, just as that year's Federal-Aid Highway Act and Department of Transportation Act enacted the first components of the Parklands Statutes. Administrators thus had to show the absence of "feasible and prudent" alternative routes before they could send I-40 through the park. While this review process was underway, however, the locals who brought suit identified two issues.[50]

First, federal funds had already been dispersed to purchase rights-of-way on either side of the park, a common if duplicitous procedure officials had long employed to secure favored routes through potentially controversial areas. By laying pavement up to the park's edges, administrators manufactured conditions in which no other route would be feasible or prudent to connect the already-completed segments. Second, despite such practices, neither Volpe nor other administrators appeared to have produced factual findings to support their choice to route I-40 through the park. Yet in subsequent litigation, the District Court and then the Sixth Circuit ruled that factual findings were not required and refused to order the deposition of Volpe or other officials regarding the route selection process. Neither court found that Volpe had exceeded his authority as secretary of transportation in routing interstate construction through parkland. But even so, the guiding statute made clear that official determination was necessary to allow parkland development to proceed. And evidence of that official determination was unclear. Nonetheless, the Sixth Circuit, as in *Ellington*, ruled in favor of the administrators.[51]

In dissent, Judge Anthony Celebrezze expressed several concerns relating to citizen participation and parkland destruction. Before his 1965 appointment to the Appeals Court by President Johnson, Celebrezze served five terms as the mayor of Cleveland. There, he had gained prominence for overseeing a $140 million "urban renewal" program—the type of program that often drew criticism for discriminatory impacts during the 1950s and 1960s, veiling racism in promises of economic growth. During these years Cleveland experienced a dramatic period of "white flight," and the city's poverty levels rose. Future archival work may reveal how, if at all, Celebrezze connected the dots among urban renewal, racism, and poverty in Cleveland. What is clear is that as mayor he was in direct administrative proximity with such issues, which he subsequently confronted on a national scale while serving as secretary of Health, Education, and Welfare in the presidential administrations of John F. Kennedy and Lyndon B. Johnson, before Johnson sent him to the bench.[52]

Celebrezze began by questioning procedure. Arlo Smith, a leader of the Citizens to Preserve Overton Park activist group, testified that there was no official statement from highway administrators attesting that Overton Park provided the only "feasible and prudent" route for interstate construction. Additionally, Lowell Bridwell, the federal highway administrator, had testified before Congress that the decision to send I-40 through Overton Park had been "completely in the hands of the city council," despite the statutory requirement that its location must be approved by the secretary of transportation.[53] In different ways, Smith and Bridwell each suggested that statutory procedure had been flouted. Furthermore, was there a factual basis for the claim that Overton Park was the only possible place to locate I-40? Celebrezze thought there was "a litany of other competent evidence that the route chosen was neither the only feasible nor prudent alternative." His dissent implied that judicial deference had gone a step too far—in both procedural and substantive terms—in the matter of I-40's construction.[54]

Moving on from procedural concerns, Celebrezze focused his attention on the values that would shape which instances of interstate construction petitioners might successfully halt in the courts. This brought due process and equal protection back to the fore. "Public parklands are the only remaining weekend sanctuaries for vast numbers of city dwellers from the polluted urban sprawl," Celebrezze wrote. "A threat to a neighborhood parkland is a threat to the health, happiness, and peace of mind of all the neighborhood people," he continued. "Congress recognized this fact. The Highway Act therefore requires that the public, and their experts, be consulted, and that their testimony be weighed in the manner of courtroom evidence by the federal officials responsible for funding highway projects." Volpe had not "fulfilled his duty" by merely "seeing that the requisite hearings are conducted and the necessary advice solicited." Instead, "procedural due process" meant that the secretary of transportation must ensure "that the people of this country be listened to, and heard, on matters affecting their daily lives as directly as the environment in which they live." Echoing the opinion in *Ellington*,

Celebrezze agreed that the courts should "never interfere" with technical expertise, but they did have a "solemn responsibility" to ensure that the technical aspects of interstate construction were "in accordance with the laws of the United States and the public welfare." Celebrezze had laid the groundwork for judicial intervention in interstate planning. His dissent also armed the petitioners with new arguments when they appealed their case to the Supreme Court.[55]

As *Overton Park* moved through the courts in 1970, agency lawyers considered its implications. Reis, the FHWA lawyer who had warned of rising judicial intervention in administrative action after *Ellington*, updated his administrative colleagues. Following the Sixth Circuit's *Overton Park* ruling, he reported that the FHWA had "prevailed," because the injunction had been dissolved. He thought the court had given "great stress to the presumed regularity of the acts of the administrator[s]." Most important, from the perspective of administrators, the court had agreed that the highway agency had given proper consideration to "alternate routes and alternate methods of construction by Federal and State officials." The judges found that highway administrators had "applied wisdom and reason," and that this exercise of expertise had been in accordance with all necessary procedures. Reis felt sure that "no court" in the future would intervene and "determine the issues" if administrators continued to show the application of reason when they selected construction routes.[56]

Reis predicted that judges would not intervene in substantive matters if they were convinced that administrators had followed administrative procedures in evaluating the substantive matters at hand. Administrators' expertise would govern. If highway administrators could "show consideration of protests" against agency action in their reasoning, Reis was confident that administrative decisions would "be upheld no matter what the personal predilection of the judge." His confidence in administrative authority had been born in the preceding decades of judicial deference to the value-laden choices of the administrative state: "The determination on the location and construction of highways is still a political one," he averred, and judges would "not substitute their judgement for that of executive officers entrusted with this decision." The manner of decision-making was poised to undermine Reis's confidence in judicial deference.[57]

While Reis prematurely interpreted the Sixth Circuit's ruling as a victory for interstate highway development—and for administrative governance in general—he identified the issue that the Supreme Court would soon take up when it evaluated the Sixth Circuit's decision. The Parklands Statutes required the secretary of transportation *not* to approve any highway construction through parkland unless there was "no feasible or prudent alternative." For highway administrators, this meant showing how the secretary of transportation had made his decision to route I-40 through parkland—exactly where it was not supposed to go, unless there was no alternative. The lower courts had agreed with administrators that agency findings did not need to be articulated in any "contemporaneous document" created during the route-planning process. Essentially, the lower courts considered the findings to be within the protected purview of administrative expertise and discretion. Reis knew administrators were on shaky ground on this point. He admitted that convincing the lower court judges of the propriety of agency findings and decision-making had been a "difficult task." He recommended that future administrators should have "a formal document prepared on all projects affecting parks and recreation areas showing the factors considered in determining that there are no feasible or prudent alternatives and that all possible planning is done to minimize harm." It was a prescient suggestion. As the Supreme Court's subsequent *Overton Park* ruling indicated, the tides had shifted against taking administrative decision-making at face value.[58]

In reviewing *Overton Park*, the Supreme Court began with the determination by highway administrators that there was "no feasible and prudent alternative to the use" of Overton Park for highway development, and that they had conducted "all possible planning to minimize harm" to the parkland. The plaintiffs, on the other hand, claimed that this determination was invalid, because the administrators had not shown how they had eliminated alternate route options. They further claimed that "the Secretary had to make the determination in writing, and with factual findings; and that the Secretary had not made an independent judgement, but merely relied on the judgement of the city officials." Administrative discretion had ruled the day—and threatened to destroy the park. The Supreme Court agreed with the lower courts that the Secretary's determination did not have to be in writing, finding no procedural requirements on this point. But the Supreme Court also found that the lower courts "were in error in deciding merely on the basis of affidavits submitted by the defendants that the Secretary's decision was not arbitrary or capricious." The claims of administrators, that is, were not sufficient to determine if procedure had been followed.[59]

The Supreme Court remanded the case, ordering the District Court to retry the case—and this time to review the secretary's determination "on the basis of the full administrative record which was before the Secretary at the time he made his decision," including public commentary. If the lower court found that "the administrative record" was "not sufficient to disclose the factors which the Secretary relied on," the court would have to hear testimony from "the Federal officials who participated in the decision." They would have to "explain their actions in court," a departure from broad acceptance of and deference to administrative expertise. If it was not a positive assertion in favor of enhanced citizen participation in the administrative process, the limited role of citizen participation displayed in the case had nonetheless helped produce the outcome.[60]

The *Overton Park* decision meant that courts would no longer take administrators at their word—as the lower courts had in the litigation. The Supreme Court still refrained from making its own substantive determination, but the decision undermined administrative authority just enough to give greater voice to citizens who challenged the substance of administrative decisions, albeit still via procedural claims. *Overton Park* opened space for litigation that dealt with procedural review yet was grounded in substantive harms. In this light, the decision may look like an attempt by the Supreme Court to tame the administrative state's nonparticipatory tendencies on behalf of citizens. And some citizens appeared to benefit from the decision: Park protectors in Memphis, in a functional sense, won the day. But this victory—also a victory for the politics of whiteness and environmentalism—did not revise the deeper problems of government that had come on display. Such a ruling could not, by its nature, affirmatively enhance community participation in the processes of state building. The deeper problems from which the case arose would continue to define the interstate highway system's social outcomes. The legal landscape had changed just slightly, and not for everyone.[61]

*

Ellington and *Overton Park* arose from protests over immediate harms as citizens went to court in hopes of preventing interstate construction from destroying their communities and parks. From those local contexts, lawsuits probed far deeper into the practices of modern American state building than freeway-fighting litigants may have foreseen. The cases illuminated perilous governmental mechanisms that prefigured inequitable construction choices and drove the interstate system's destructive outcomes. Together, *Ellington* and *Overton Park* put racist and classist official priorities on display, while revealing tensions between administrative authority and participatory democracy. Inseparable, if abstract, questions about the roles of citizens, legislators, administrators, and judges in the work of state building hung in the balance.

In more straightforward terms, the 1967 *Ellington* decision upheld a vision of administrative authority that, by the time the Supreme Court denied certiorari, was under sustained popular and political assault from all points of the political spectrum. Just a few years later, in 1971, the *Overton Park* decision established a presumption of judicial review of governmental action.[62] That presumption, encompassing expanded standing for litigation, appeared to give voice to citizens in a more direct manner than they might have found through legislative representation. But *Overton Park* also allowed the administrative state to retain much of its authority. The root problem with which the Supreme Court took issue in the case was the absence of administrative justifications for actions planned and taken—justifications demanded by citizens through their lawsuit. And the Supreme Court insisted on the need for those justifications by invoking the Administrative Procedure Act as well as the Parklands Statutes, to which citizens could point as they demanded an explanation for the destruction of Memphis's green space. The *Overton Park* decision enhanced judicial review of administrative action and expanded opportunities for citizen participation through litigation, while constraining both aspects of administrative oversight to close statutory adherence. While judicial review and citizen participation had been overriding issues as the litigation moved through the courts, it was on statutory matters that *Overton Park* focused the attention of administrators and their lawyers after the Supreme Court's ruling. The decision stands out in administrative law for its reformist inclinations, yet it did not refashion democracy.[63]

Overton Park's outcome meant that citizen participation could, in some instances, now present a roadblock to interstate construction. The key for administrators—as agency lawyers like Reis had long made clear—was to preemptively identify the substantive issues on which citizens would now find widened standing to sue. *Overton Park*, in the end, enabled sharply limited participatory opportunities for citizens in the work of administrative state building. Participation was only possible on the terms set forth by the Parklands Statutes and the emerging tenets of environmental law that suffused the case. It was to this legal area that administrators and their lawyers shifted their focus in *Overton Park*'s aftermath, while construction in neighborhoods proceeded to produce discriminatory outcomes nationwide. *Overton Park* revealed that the courts would insist on "strict compliance" with environmental laws, according to Lloyd Reeder, another FHWA lawyer tasked with updating state and federal agencies on legal changes in 1971. On this point, he felt that "the best advice" he could offer his administrative colleagues was to "anticipate problems" related to "hearings, civil rights, and the whole environment." But based the outcome of *Overton Park* and his summary of developing case law, it was clear which of these issues would take precedence.[64]

Administrators did not have carte blanche to destroy communities while protecting parks, but the calculus was plain. Legal barriers to pushing interstate highways through neighborhoods were lower than legal barriers

FIGURE 23. A bulldozer pushes through a North Nashville neighborhood to make way for Interstate 40, 1968. Jack Corn / USA Today Network via Imagn Images.

protecting parks. As Reeder summarized, the best bet for administrators was to give the protection of parklands "paramount importance" when selecting interstate routes. This meant that the legal framework that emerged in the wake of *Overton Park* was poised to worsen the effects of interstate highway construction in urban communities, exacerbating rather than reducing inequities. And even when it came to parkland, citizens had limited possibilities for participation. If administrators were mindful of strict compliance with statutory guidance, they would have little trouble with the courts on environmental issues.

Even if protecting parkland was now of "paramount importance," the results of *Ellington* and *Overton Park* indicated the continued power of administrators to build how they pleased. The outcome also reflected the significance and endurance of what historian N. D. B. Connolly calls "Jim Crow liberalism." Judges and officials made way for the generally white preoccupation with park protection, while articulating "opaque" legal reasons to ignore the generally Black preoccupation with community protection. Of course, the white concern with parkland and the Black concern with community life were born of the blunt fact that white neighborhoods were rarely targeted for interstate highway development. This discrepancy was justified with the financial calculus of buying the cheapest available land, which meant land inhabited by poor communities of color. "Letting the 'market' decide," writes Connolly, was itself a method for ensuring the preservation of "white power" and its priorities.[65]

Such discriminatory priorities and practices had long been etched into the American landscape. But at the beginning of the 1970s, for a brief moment, other possibilities seemed open. Reformers might have hoped that the *Ellington* decision was a remnant, a vestige of past lawmaking and jurisprudence that had been overly deferential to the administrative state at the cost of participatory and even representational democracy—and insufficiently protective of Black citizens' substantive concerns and civil rights. There were reasons for such hope. After all, the same term that the Supreme Court decided *Overton Park* saw the justices reach two separate decisions, both unanimous, that struck significant blows against racial discrimination and segregation. The first was *Griggs v. Duke Power Co.*, holding that Title VII of the Civil Rights Act of 1964 prohibited employment practices that had disparate impacts on Black workers, regardless of employers' intent to discriminate. The second was *Swann v. Charlotte-Mecklenburg Board of Education*, holding that busing was an appropriate remedy to school segregation. But a few years later, the 1974 *Milliken v. Bradley* decision saw the Supreme Court hold that school district lines could not be redrawn to combat segregation unless there was evidence of discriminatory intent. Two years after that came *Washington v. Davis*, holding that evidence of discriminatory intent was a prerequisite to finding governmental discrimination, even if a policy had disparate impacts on protected groups. *Ellington* and *Overton Park*, in retrospect, helped pave the way for such developments by producing decisions that delimited the presence of substantive claims, the recognition of civil rights harms, and scope of participation—adding further durability to "Jim Crow liberalism" and its cascading forms of segregation.[66]

In sum, *Ellington* and in *Overton Park* reinforced a legal bulwark that protected the advantages of white, affluent litigants who wished to preserve unbuilt spaces and blockaded Black litigants who wished to preserve their neighborhoods. The cases' wide disparities and narrow reforms combined to insulate and maintain the paradigm of infrastructural state building responsible for the benefits and burdens of the interstate highway system. That paradigm seldom impacted all equally as it bulldozed local prosperity in the name of national economic growth, cementing racial and class hierarchies with all the permanence of asphalt and concrete. The legacy of this state-building regime endures: The concrete leviathan that defines the modern American state's territory is a creature of law that has inscribed spatial inequalities at dramatic physical and temporal scales. Any possibility of reform requires communities to do nothing less than remove the highways whose construction was impossible to avert in the first place—and even that does not solve the deeper problem of democratizing the administrative foundations that undergird infrastructural development.

6

Shaken to the Foundation

In April 2019, something remarkable happened in Syracuse, New York. The state transportation department, working with local and federal counterparts, agreed to tear down a section of Interstate 81 running through the middle of the city. The primary target was a 1.4-mile-long elevated viaduct carrying four lanes of traffic. In the 1950s, officials had conspired to route this viaduct directly through the predominantly Black 15th Ward, at the city's geographic core. The neighborhood was home to about 90 percent of the city's Black population—some ten thousand people. Throughout the 1960s, the viaduct's construction displaced as many as 1,300 of these residents, and other infrastructure projects connected to interstate development soon pushed that number up to 2,500. Officials seized homes and took aim at many of the area's most successful businesses, leveling nine out of every ten buildings in the 15th Ward. The area's remaining residents crowded into places like the Pioneer Homes, a housing project located just feet from the viaduct.[1]

In place of the vibrancy and community that the 15th Ward's homes and businesses once offered, a steel, asphalt, and concrete edifice rose up several stories—a barrier cleaving through the city, with a few mostly unwalkable roadways snaking through the dim recesses beneath the noise, weight, and oppressive exhaust of heavy traffic. The I-81 viaduct destroyed the Black business district and ably separated it from Syracuse's whiter and wealthier areas. In an era marked by *Brown v. Board of Education*, construction plans like this, in cities across the country, created barricades to racial integration as de jure segregation fell away.[2]

With the arrival of the viaduct removal plan in the twenty-first century, Syracuse fashioned itself into a national leader in redressing the interstate highway system's racist and classist harms. In place of the viaduct, the city and state would collaborate to build a "community grid." Expected to cost about $2 billion in state money, this amounted to reconnecting and improving the network of city streets that I-81's construction had destroyed some fifty years earlier. Once completed, the project would enable a productive, livable, and walkable urban community at the city's center. Interstate 481—a beltway around the city—would be redesignated as I-81 and updated to accommodate more traffic. The community grid plan devised a blueprint, at least in theory, for rebuilding some of what had been lost, and rerouting I-81 on an existing highway minimized the likelihood of simply shifting highway burdens from one location to another—there would be no need to bulldoze a fresh path and impose new harms.[3]

Syracuse's mayor, Ben Walsh, was an important supporter of the community grid plan. He framed it against I-81's history. His "greatest concern" was for the people living in the 15th Ward, he said in 2021. "They have borne the biggest burden in the past, and their neighborhoods must be protected and invested in," he wrote.[4] Those residents, as it turned out, had serious concerns about the possibility of the community grid spurring gentrification, raising rents, and causing a new wave of displacement—issues that drew increasing focus from local activists and organizations like the New York Civil Liberties Union (NYCLU).

But Walsh's willingness to express concern for the community, however anodyne and imperfect, reflected how far Syracuse had come. Walsh's grandfather, William Walsh, had served as Syracuse's mayor at the time of the viaduct's construction. When the Congress on Racial Equality (CORE) protested the viaduct in 1963, the elder Walsh announced that he would "not accept" their actions and rapidly arrested ninety-eight picketers. That show of force was only the latest in a string of episodes that saw city officials censor outcry over conditions in the 15th Ward, use military tanks to demolish part of the neighborhood to make way for "urban renewal," and finally turn to the interstate highway project to destroy the community once and for all.[5] The city's searing postwar racism became mirrored by its tentative position at the vanguard of a growing twenty-first-century movement to revise racist and economically inequitable infrastructure.[6] Repairing the urban space the viaduct was built to shatter, the community grid aimed to help reverse generational declines in wealth, opportunity, and well-being. Those issues had compounded for decades, dating back to the origins of the interstate highway project.

Tracking the I-81 viaduct's life and death in Syracuse underscores the necessity of linking past choices with contemporary realities in efforts to bring about infrastructural reform. Understanding the 15th Ward's history over the past eighty years is key to understanding the full scope of the area's devastation and disenfranchisement today. That history is also key to evaluating the promises and perils of the community grid plan—as an end in itself, and as a

FIGURE 24. The Interstate 81 viaduct takes shape in the middle of what remains of the 15th Ward in Syracuse, New York, 1965. Courtesy of the Onondaga Historical Association.

model for infrastructural justice efforts in the future. To repair Syracuse's city center is a momentous task; to revise the deeper governmental problems of democracy, participation, and equity that gave rise to the interstate highway system's worst features in the first place is an even more challenging endeavor.

*

During World War II, while federal officials were busy crystallizing plans for the interstate highway system, Syracuse became a model for American urban development. In 1942, *Fortune* magazine—looking ahead to the end of the war—published a series of pamphlets dedicated to "comprehensive economic planning."[7] The series was an expression of the era's high modernist impulses and reflected a desire to put the nation on strong footing through careful political and economic design. *Fortune*'s editors began by calling for planning efforts capable of achieving John Maynard Keynes's vision of a nation-sized "economic system" whose function and success would be measured by tabulating "national income."[8] The resulting arrangements, *Fortune* explained, would be complex and interrelated, involving everything from agriculture to utilities to public works. *Fortune* dedicated an entire section to improving transportation systems, key to integrating a national web of productivity.[9]

Fortune's efforts were detailed but abstract, and the magazine's editors soon searched for a metropolitan area where they could take their proposals and test them out in "concrete terms."[10]

They settled on Syracuse, home to 206,000 inhabitants—the forty-first largest among the nation's hundred largest cities. It was a perfect "middle-sized" metropolitan area to study, near the median yet slightly aspirational, as well.[11] Syracuse was already a robust industrial center, thanks to over a century of development that flowed from its location along the Erie Canal and major rail lines. World War II ramped things up further. Most of the city's factories received Army and Navy "E's"—for "Excellence in Production"—during the war. After the war's end, Syracuse's manufacturing sector was so well diversified it actually was not especially representative of urban America: The county was one of only thirty-three in the entire nation that had representation in all twenty of the census's industrial group categories.[12] To the editors of *Fortune*, Syracuse was the beacon American city, poised to lead other metropolitan areas to still higher heights of postwar productivity and growth.

Fortune funded five city planning consultants to set up shop in Syracuse, where they helped establish the "Syracuse-Onondaga Postwar Planning Council."[13] The council was composed of various leaders from Syracuse University, the business community, and the city and county governments. Its goal was simple: to realize the "high economic potential" of the city and surrounding Onondaga County. The council worked to identify obstacles to Syracuse's upward trajectory, making particular note of "blighted areas" and the city's "inadequate" automobile infrastructure.[14] The city would need "a strong controlling authority" to rectify these problems.[15] Once the issues were resolved, however, *Fortune* was confident that Syracuse would be a "success story," showing off what "scores of American cities" could do if they engaged in long-range economic planning. It was a city to be "envied," *Fortune* claimed—and hopefully emulated.

At the center of the shining city *Fortune* depicted was the 15th Ward. Its small size—about a half a square mile—contributed to the sense of community residents identified as one of its most important characteristics.[16] In the 1930s, local newspapers ran features with titles like "No, It's Not Harlem—It's a Syracuse Block Party." In the depths of the Depression, cultural expression blended with material aid: That block party saw crowds of smiling people dancing to jazz music at one of the area's community centers, their cover charge benefiting the "Negro Boy Scouts." Many businesses had generations-deep roots in the community, offering economic and social structure and providing for the community's needs—shops like Williams Bros. Grocery, Schor's Meats, and Aunt Edith's Luncheonette.[17] Aunt Edith was known for

FIGURE 25. Williams Bros. Grocery in the 15th Ward, Syracuse, New York, date unknown. Courtesy of the Onondaga Historical Association.

FIGURE 26. Schor's Meats on Harrison Street in the 15th Ward, Syracuse, New York, 1963. Courtesy of the Onondaga Historical Association.

FIGURE 27. Patrons smile in front of Aunt Edith's Luncheonette on Harrison Street in the 15th Ward, Syracuse, New York, 1951. Richard Breland / Onondaga County Public Library.

refusing service to teenagers who skipped school, just one example of the watchfulness and care that appeared to infuse the community.[18] Clarence Dunham, a 15th Ward resident, recalled what the area was like then: "Most of the families knew everybody. If they didn't know you, their grandparents did. Or they knew your mother, your father, your uncles. It was very tight-knit, and it was lovely."[19]

But the 15th Ward was also a community shaped by racism and disenfranchisement at the hands of the city's white elite. A 1937 Home Owners' Loan Corporation "residential security map" lined nearly the entire 15th Ward in red ink. As in countless Black neighborhoods across the nation, appraisers used this "redlining" to label the area as "hazardous" for the subsidized mortgages that were a ticket to prosperity for the era's white Americans. The simple mechanism abetted cyclical disinvestment and poverty along racial lines. Active segregation efforts prevented Black Syracusans from moving into other parts of the city. Within the 15th Ward, the absence of capital for expanding or enhancing the existing supply of housing ensured a captive market. Landlords let buildings fall into disrepair, overcrowded their tenants, and gouged rents.[20] In a landscape defined by exclusion, the community had few levers of change to grasp. In 1935, a city charter had revoked the 15th Ward's political representation on the city council. A decade later, the Syracuse-Onondaga Postwar Planning Council routed their hoped-for modern highway directly through the 15th Ward. Two years after that, in 1947, the first official map for the interstate highway system located Interstate 81 along the same path.[21]

*

By the late 1940s, the racist skew of Syracuse's postwar planning efforts was readily apparent. City leaders believed highways would offer a path to continued economic growth, while simultaneously providing a means of wiping out a redlined Black neighborhood now castigated as a "slum."[22] The Great Depression and the war years had consumed Syracuse. When peace came, waves of students, many fresh from fighting fascism overseas, arrived at Syracuse University on the GI Bill. The end of World War II also brought energy into the Black freedom struggle nationwide, and community members in the 15th Ward began to give louder voice to the discrimination and deprivation they faced.[23] Their efforts were further amplified when student activists began to join in. The first flashpoint came about almost by accident. It quickly illuminated not only the problems of the 15th Ward but the limits of midcentury American democracy, which became harshly apparent as the postwar infrastructure boom and the civil rights movement took off alongside each other.

The trouble began a few minutes before 7 p.m. on March 8, 1949. It was a raw Friday evening in a city known for foul weather—following an afternoon of freezing rain, the temperature had plummeted to around twenty degrees, bringing snow with it. But a crowd of about eighty people had gathered at the corner of Harrison and South McBride in the 15th Ward. A twenty-four-year-old Syracuse University student named Irving Feiner held their attention with a loudspeaker, delivering an important message: The location for a talk later in the evening by O. John Rogge had changed location, from a nearby public school auditorium to the Hotel Syracuse.[24]

The Syracuse chapters of the American Labor Party (ALP) and the Young Progressives of America (YPA) had sponsored the talk. Feiner was a member of both organizations. The ALP had become especially attentive to the 15th Ward, laying the problems of dilapidated, exploitative housing and schools that had become "fire traps" at the feet of officials.[25] Hosting Rogge would bring further attention to the cause. Rogge had held several positions during the Roosevelt administration, including US assistant attorney general. In the intervening years he had emerged as a prominent civil rights advocate. By the time he came to Syracuse, he was in the middle of appealing the case of the Trenton Six, a group of young Black men found guilty by an all-white jury of killing a white shopkeeper. The case, rife with official misconduct and dishonesty, had captured national attention.[26]

Feiner was agitated as he announced the venue change for Rogge's event. It had not been by choice. At the last minute, city officials had revoked the permit for the school auditorium, quashing the opportunity to bring a nationally renowned civil rights lawyer into the 15th Ward to engage with the community.[27] Feiner, born in the Bronx to Polish parents, had dropped out of high school to serve in World War II. The GI Bill made higher education possible for him, and he intended to become a civil rights lawyer himself.[28] By age twenty-four, his life experiences had already furnished him with clear political principles, and he had no patience for the games the city was playing with the ALP's and YPA's efforts to organize and educate in the 15th Ward. After telling the crowd to go to Hotel Syracuse instead of the school auditorium later on, he launched into a tirade against the city establishment that had attempted to censor the event. The mayor was a "champagne-sipping bum," and his cronies were no better than the "Nazi Gestapo" he had fought overseas, he thundered into his loudspeaker. After about twenty minutes in this vein, he said something else, the exact wording of which would become much debated. Most likely, he invited the crowd to "rise up and fight for their rights by going arm and arm to the Hotel Syracuse, black and white alike, to hear John Rogge."[29]

At this point, two police officers had arrived to surveil the crowd, and they claimed to hear a man say, "If you don't get that son of a bitch off, I will go over and get him off there myself." The officers proceeded to ask Feiner three times to cease speaking; he refused, and they arrested him for disorderly conduct.[30] He quickly posted $50 bail and made it to the hotel in time for Rogge's talk. A crowd of about two hundred had gathered, and he kicked things off by recounting his arrest. When Rogge took the stage, he called the night's events "fascism—American style."[31] The next morning, Feiner pleaded not guilty, and his bail was reset at the extraordinary amount of $1,000. Even local newspapers, which were not sympathetic to him, were alarmed by the significance of this, calling the sum "unprecedented" and noting that criminals and racketeers usually faced no more than $250 bail.[32]

Feiner's case got underway. The district attorney as well as the local judge, William Bamerick—who made remarkably little secret of his presupposition of Feiner's guilt—must have known they were on shaky ground. The authorities grudgingly recognized that Feiner had a "right to hold a street meeting at this locality, to make use of loudspeaking equipment in giving his speech, and to make derogatory remarks concerning public officials."[33] So the prosecution set its sights on the content of what Feiner had said, with some curious editorial changes. Witnesses claimed that Feiner had called on the crowd to "rise up and fight for their rights by going arm and arm to the Hotel Syracuse." At

FIGURE 28. Irving Feiner, center, at the O. John Rogge event, 1951. The image was captured shortly after Feiner's arrest. At right is Nate Zeppetello. The figure at left is likely Mary Ann Zeppetello. Both Zeppetellos held leadership positions in the Onondaga County Communist Party and had stood with Feiner as he was arrested. *The Post-Standard* / Syracuse.com.

some point during the initial trial this became "rise up in arms and fight for their rights." The addition of "in arms" came courtesy of the arresting police officers, who had not used the phrase in their initial statements.[34]

This inkling of violent intent was all Judge Bamerick needed. He had already been able to block a jury trial, and he quickly pronounced that Feiner had acted "outside the law in setting race against race."[35] He told Feiner that his activities had not been "acts of good citizenship," stated that Feiner had been "deliberately agitating and goading police officers" into arresting him, and sentenced him to thirty days in jail.[36] Syracuse's primary newspaper, *The Post-Standard*, claimed the case would "soon be forgotten" and hoped that the jail sentence had made Feiner a "wiser and more mature young man."[37] The city was trying to sweep it all away. "Democracy," the ALP announced, had been "thrown down the sewer" in Syracuse.[38]

Feiner, however, was not finished yet. He appealed his case all the way to the Supreme Court. The American Civil Liberties Union (ACLU) led his defense, with direct input from Arthur Garfield Hays, the organization's cofounder, who had once defended Sacco and Vanzetti.[39] *Feiner v. New York* became an infamous case, a landmark ruling that curtailed the right to free speech. Chief Justice Frederick Vinson, writing for a 6–3 majority in 1951, agreed with Syracuse police and Judge Bamerick that Feiner's speech had created a "clear and present danger of riot." Therefore, the "interest of the community in maintaining peace and order" trumped the exercise of free political speech calling for equal rights.[40]

In a searing dissent, Justice Hugo Black wrote that the decision marked "a dark day for civil liberties." He decried the Supreme Court majority's willingness to ally with a local trial court judge who had "accepted the testimony of the prosecution witnesses on all important points." His colleagues had "made a mockery of the free speech guarantees" of the Constitution. Black unwound the flawed logic of the majority opinion, which went like this: Violence was about to break out, the police were therefore justified in demanding Feiner stop his speech, and they were further justified in arresting him when he did not. Nowhere in this argument was there any question of the legality of Feiner's presence on the street corner, nor the content of his speech. Did not the police have a constitutional obligation to protect Feiner and his speech from the violence and censorship of a supposedly hostile crowd? At the very least, why did they not arrest the man who made an actual threat of violence against Feiner—a man the prosecution had never identified? It was not Feiner who was the problem; instead, Black indicted the police, who acted in "deliberate defiance" of their official duty. Erring, too, were the lower-court judges and the authors of the Supreme Court's majority opinion, all of whom accepted the prosecution's story in order to ignore calls for racial equality and democratic government. In doing so, they had invented "a simple and readily available technique" with which "minority speakers can be silenced in any city."[41]

Justice Black's dissent underscored another facet of the case that had received little attention. The intersection of Harrison and South McBride, where police arrested Feiner, was a common meeting place in a community that had few public spaces. There was a standing Friday evening assembly there. That was why Feiner had picked it. There had never been any problem until the topic of the unofficial weekly town hall turned, that March night, to civil rights and the Black freedom struggle.[42] Before another decade passed, the city marked the intersection for demolition. Exit 18 on the southbound side of the I-81 viaduct would soon come swooping down where Feiner had stood.[43]

In 1949, it was possible to foresee that looming neighborhood destruction. Congress was debating the Housing Act of 1949, which would provide federal funding for wiping out "slum" areas like the 15th Ward, ostensibly to make way for more habitable development but, in reality, destroying far more than it ever built.[44] In a short time the interstate highway project would provide even better funding for such "urban renewal." Feiner and other activists could not avert the 15th Ward's destruction via such policies, and Feiner paid a particularly steep price for his efforts. By the end of his jail time, Syracuse University had expelled him, and several law schools rescinded admissions offers. (Syracuse University finally readmitted him in the 1980s to finish his degree, by which time he was a successful small business owner. And in 2006, in the height of irony, the university invited him back to campus to inaugurate a new center devoted to free speech.)[45]

Feiner's activism connected the granular realities of American inequality with the most rarefied forms of national lawmaking. His Supreme Court case revealed the steps by which the United States might slip into "totalitarian control," as Justice Black wrote.[46] The limits of democracy were at stake, and the conceptual and societal import was clear in the majority opinion and the dissent. All of that arose from Feiner speaking off the cuff, lambasting the racist city elite through a tinny loudspeaker on a cold March night long ago. The 15th Ward was suffering from the choices of a racist, classist, undemocratic city, and he aimed to make it as hard as possible for Syracuse's officials to excuse what they were doing.

Activism in the 15th Ward appears to have tapered off for a time after Feiner's arrest and imprisonment. It is easy to infer the connection—the case must have had a chilling effect on the city's residents. Feiner was white, a veteran, a college student. Protestors without those advantages might fare even worse. And there was the matter of his absurd bail. As a letter to the editor of *The Post-Standard* put it, "Will it ever be possible from now on to make a speech criticizing the city administration without risking arrest? Will a Syracusan ever again be able to speak out loud in opposition to authority without being prepared to post $1,000 bond?"[47] That the paper even published the letter, however, suggested a glimmer of hope. It is difficult to gauge the direct impact of Feiner's arrest and court case on the public's interest in the 15th Ward, although it seems certain that less attention would have followed if the police had simply let Feiner speak.

The 15th Ward continued to draw notice as the civil rights struggle picked up steam nationwide. Even *The Post-Standard*, which had repeatedly disparaged Feiner, began to take a freshly critical look at Syracuse government in the years ahead, trying to suss out who or what was to blame for the poverty and segregation that defined the city.

*

In 1954, three years after the *Feiner* decision, a thirty-one-year-old reporter named Walter Carroll published a ten-part series on conditions in the 15th Ward in *The Post-Standard*. Carroll was new to Syracuse—he had joined the paper in 1952—and perhaps an outsider's perspective helped him clarify the complex story. He spent several months in the 15th Ward, meeting residents and discussing what caused the area's severe poverty. Accompanying him was a young welfare caseworker named John A. Williams who, in a few years' time, would emerge as a critically acclaimed writer. Born in 1925, Williams grew up in the 15th Ward. His early work dramatized the community's plight and eventually lifted far broader social analyses from the experiences he had there. For now, though, he served as a guide and interlocutor for Carroll. The two shared meals with residents, drank in the neighborhood's illegal bars, and immersed themselves as best they could. Williams later opined that Carroll's reporting "shook Syracuse to its foundations." His ten-part series blamed the 15th Ward's abject poverty not on the community's residents, but on the city's government.[48]

Carroll wrote for the sizable Syracuse audience that never ventured near the 15th Ward, letting their racism and entwined belief in the area's danger demarcate a firm boundary line around Syracuse's geographic center.[49] For such readers, Carroll narrated a detailed tour. He described a typical dwelling, "a drafty frame structure" shared by six families, one family to each of the building's rooms, "one filthy kitchen, one filthy toilet" shared among them. Running water was a rarity in the neighborhood, and "thousands of roaches seethed over the walls" of each home he visited. As he spoke with one family, he watched as six crawled from the pillow of a twelve-year-old boy. In the hallways, he heard "rats scurrying through the walls." In pantries, he found that the rodents chewed through the packaging of any food items left unsecured. In one kitchen, he noted a gallon jug half filled with a mysterious orange liquid, two inches of dead roaches floating on its surface.[50]

Syracuse's health commissioner blamed such conditions on a "lack of pride" among the 15th Ward's residents.[51] Carroll insisted other forces were at play. The 15th Ward's residents did not live this way by choice. Elite white Syracusans had long maintained that the area's conditions could be explained only by "desire" on the part of the community—a myth that covered a multitude of governmental and private-sector sins.[52] Carroll countered that "the problems in the 15th Ward have their roots in prejudice."[53] That prejudice, he explained, was baked into the city's design.

Carroll paid particular attention to Syracuse's housing market. His seventh column in the series, titled "Negroes Denied Homes Except in Ghetto Sector," identified a long-standing alignment between discriminatory city policies and real estate industry practices. One of Syracuse's "leading real estate men"—who was not willing to be named—told Carroll that he was "against Negroes living in white neighborhoods. The value of property goes down when they move in. . . . We definitely do not sell to them." Carroll prodded: If the real estate market's brokers would not sell to Black Syracusans, what should be done about the 15th Ward? "You've just got to accept it," the man replied. Another real estate agent explained that "Negroes are just not shown houses for sale." At the time of Carroll's reporting there were about seven hundred listings throughout Syracuse. There was no "out-spoken discrimination," the agent explained, about refusing these to the city's Black residents. Instead, it was a "subtle thing." Agents showed them properties only "in the slum area" and told them "that's all there is." If they sold a Black family a home anywhere but the 15th Ward "there would be the devil to pay."[54] Segregation and poverty occurred by design.

Quoting a local Black pastor, Carroll concluded his ten-part series by stating that it was time to end the "constant humiliation" that Syracuse imposed on its poor and nonwhite population. He hoped the city would invest in low-cost housing development and other infrastructure projects that might advance "equality of opportunity" in the urban landscape. He told his readers once again that Syracuse's Black population lived in the poverty of the 15th Ward "because they cannot live elsewhere."[55] And he even reminded Syracuse that eighty years before, many of its white residents had joined the Union Army and died on battlefields in the South, fighting "to end slavery and for the equal rights of Negroes." Times, evidently, had changed. "The spirit of what they fought and died for is not dear to our hearts," he lamented.[56] Beyond these prods, he offered little advice on how to overturn the city's racist paradigm. The best he could do was suggest the creation of an "inter-racial clinic" to help Syracusans resolve their prejudices. Taking the form of an extragovernmental community group, perhaps it could "deal with the ills of racial tensions" and "improve the social health" of the city. Maybe it would provide the basis for citizens to come together and "analyze the problems of the community" in a "democratic manner."[57] It was a nice idea.

In the end, Carroll concluded that all he could really do was give Syracuse the "facts" and hopefully lay bare the racist and classist myths that elite white Syracusans draped over the city's poorest sections. It was up to Syracusans to take "action" on those facts.[58] And take action they did, for Carroll had inadvertently given city officials an excuse for "slum clearance" rather than an inspiration for investment in the community. In less than two years' time, the interstate highway project would present an even grander excuse. In the meantime, Syracuse revved the engines of urban renewal.

*

"Tanks Poised for Blitzkrieg in 15th Ward," blared a *Post-Standard* headline on August 10, 1954. It was a striking choice of words. Invoking the German "lightning war" strategy suggested a brutal surprise attack. World War II was less than a decade in the past; to use such language was a calculated decision. The paper appeared to be proud of the effort underway, celebrating the men of the Marine Corps Reserve's Second Tank Battalion—stationed at the nearby Camp Drum military base—who volunteered for demolition service. They treated it as a public relations stunt, declining to charge the city for their work as they deployed for 1104 and 1106 East Washington Street in the 15th Ward—four blocks from the public school where the city had barred O. John Rogge from speaking about civil rights a few years earlier.[59]

Walter Carroll reported on the blitz. His column—"47-Ton Punch Flattens 15th Ward Structures"—began with a haunting photograph of a man sitting on the uneven steps of the dilapidated home at 1106 East Washington. Robert McCloud's hands were folded in his lap, his eyes not quite meeting the camera, his mouth set at an angle just shy of despair. But he sat upright, his shoulders square; a neat cap on his head helped him look dignified in a filthy pair of trousers and work boots. The city had condemned the building eight years previously, likely for a mix of reasons having to do with real safety concerns and to create the eventual pretense for its removal. McCloud, his wife, and two young children had arrived sometime during that eight-year interval and lived in a back room.[60] McCloud had been able to find work two days a week as a laborer, but nothing more.[61] In May 1954, a few months before the tanks rolled in, the McClouds' presence in the condemned dwelling had become known, and the police had given them forty-eight hours to leave the city. Evidently, they had remained up until the tanks flushed them out. Now they had no choice but to move; Carroll did not say whether they had another home to go to.

Images from the day the McClouds left show a massive forty-seven-ton "tank retriever"—an oversized tank fitted with .50- and .30-caliber machine guns as well as a crane and bulldozer blade—rolling across what remained of the home. A crowd gathered around Mayor Donald Mead to watch the spectacle.[62] The marines needed less than an hour to flatten the dwelling; Carroll was amazed at how rapidly the massive tank turned it into "a cloud of dust and splinters."[63] The tank retriever was an imposing vehicle, heavily armored and designed to traverse enemy territory and rescue other tanks halted by mortar fire. Its operators looped a cable around the buildings in the 15th Ward; with one pull from the crane, it "knifed" the structures down at their base.

FIGURE 29. Robert McCloud sits on the steps of 1106 East Washington Street in the 15th Ward, Syracuse, New York, 1954. McCloud had been living in the condemned building with his wife and two children. *The Post-Standard* / Syracuse.com.

When the rough work was complete, the marines moved smaller tanks in to crush the remains of the buildings, likely using the M4 Sherman—light in comparison at thirty tons, but sporting significantly larger (75mm) artillery guns.[64] These wargames might have resulted from official requisitions, but the local newspapers portrayed the tank demolition as a casual arrangement made between the local "leatherneck" commander, Major Arthur Hale, and city officials.[65]

A month later, *The Post-Standard* noted that the railroad police had picked up Robert McCloud on the spurious Jim Crow charge of "loitering." Perhaps he had been wandering through the 15th Ward looking for a place to live. A judge added a probation violation based on his failure to leave the city

FIGURE 30. US Marines use a forty-seven-ton tank retriever to knock down the building at 1106 East Washington Street in the 15th Ward, Syracuse, New York, 1954. Mayor Donald Mead and a group of officials and onlookers observe the destruction. *The Post-Standard* / Syracuse.com.

the previous May, and sentenced him to six months in county jail. There was no mention of what became of his family.[66]

The poverty of the McClouds' surroundings must have been severe at 1106 East Washington, a condemned building that presumably had no heat, electricity, or plumbing. Still, they had made do there for months or years, perhaps with the help of alert welfare workers like John A. Williams and certainly with the aid of the community's strong culture and social ties. What little that had amounted to—a bare hope of keeping the family together and making something of life in the 15th Ward—came crashing down in an instant. This was a snapshot of slum clearance in 1950s America, enacted by force and by legal scheming and reflective of a situation in which "urban renewal" meant nothing other than "Negro removal," as James Baldwin put it.[67]

*

Later, John A. Williams reflected on the period in which he worked for the welfare department and watched as officials started to demolish his 15th Ward community. In those years he was "drowning, really, really headed for the bottom." Writing became his "lifeline" as he watched the brutal priorities of infrastructural development take hold.[68] In college, he had focused on poetry; as he turned to writing novels, one of his first animating interests was Black family life. This was the central area of exploration in his first major work, *Sissie*, which he began drafting in 1960. "That black people in this nation have been capable of maintaining *any* family structure at all is so incredible a feat that it cannot yet be measured," he reflected. He drew a straight line from the 15th Ward's conditions—and the fate of families like the McClouds—back to American chattel slavery, indignant that "the ignorant" continued to blame Black Americans for "behavior patterns" that resulted from white supremacy. Another contrasting interest for Williams was the Black experience of "guilt because of survival."[69]

Williams's own childhood in the 15th Ward had been poor but relatively stable compared to what he saw around him. His family lived first on Washington Street. His parents had arrived from the South—his father, bleeding from a pitchfork wound and hidden in a hay wagon, as a lynch mob chased him. Williams credited his mother with encouraging him to become a writer.[70] He joined the Navy in 1943 after finishing high school and served for three years as a medic in the Pacific theater. After receiving an honorable discharge, he returned to Syracuse in 1946 and enrolled at Syracuse University on the GI Bill. He excelled as an English major, corresponding with William Carlos Williams, Langston Hughes, and other luminaries as an undergraduate. He and Irving Feiner were part of the same class year, and likely met. Williams was cagey about his connection to the YPA and other leftist groups but spoke at their events on more than one occasion. During his college years and after he graduated in 1950, he worked jobs throughout Syracuse's postwar economy. As a hospital orderly, supermarket clerk, elevator operator, foundry worker, and eventually county welfare caseworker, he attained a stark understanding of life in the city.[71]

Soon after the tanks rolled through the 15th Ward, Williams moved to New York City, where his political outlook took further shape alongside his writing career. He worked as the public relations director for the American Committee on Africa, which supported African liberation movements. In 1958 he became the European correspondent for both *Jet* and *Ebony* magazines. In the early 1960s, he reported for *Newsday* from Africa. He traveled the continent widely, spending time in Ethiopia, Egypt, the Sudan, and Nigeria. In Lagos, he met Malcolm X, who had read and admired *Sissie*.[72]

That first major novel, completed in 1962 and published the following year, received acclaim. A deep portrait of a Black family, *Sissie* explored life in a fictionalized neighborhood bearing a distinctive resemblance to the 15th Ward. Some of *Sissie*'s characters stared longingly up at the university on the hill above, wondering if any of their children might go on to become a "rare black face in the crowd of students."[73] Tracking the Joplin family through time in a complex narrative composed of four flashbacks, the book was a critical success, celebrated for its ability to "chronicle Negro life in transition."[74] That transition stretched from World War II through the political and cultural shifts of the 1960s, as the Joplin family was torn apart by worsening poverty despite the era's promises of equality and opportunity. It was Williams's bad luck that *Sissie* was published almost simultaneously with James Baldwin's *Another Country. The Saturday Review of Literature* opined that Williams's work was "by far better" than Baldwin's, celebrating the timeliness of its "quiet anger that builds and builds inexorably."[75] Apparently, one literature scholar lamented, "American critics and readers could not embrace two big novels about the black experience at the same time."[76] *Sissie* never attained commercial success matching that of Baldwin's work.

Williams's next book, *The Man Who Cried I Am*, appeared in 1967 and was his most commercially successful. *The New York Times* called it "a compelling novel, gracefully written, angry but acute."[77] With anger again a prevalent theme, the novel evaluated the preceding thirty years of American history, narrated by a dying Black writer. One prominent storyline was the CIA's assassination of "Minister Q"—a fictional Malcolm X. Williams was not impressed with the sweep of time he studied in the novel, nor the supposed progress those years contained. His next major work, *The King God Didn't Save: Reflections on the Life and Death of Martin Luther King, Jr.*, appeared in 1970. It was a scathing critique of King's efforts and landed poorly so close after the civil rights hero's murder. Williams accused King of being a dupe for white interests, bought off by useless concessions and naive about the structure of power in American society. "He did not understand that it had armed him with feather dusters," Williams averred. "He was a black man and therefore always was and always would be naked of power, for he was slow, indeed unable, to perceive the manipulation of white power, and in the end white power killed him."[78]

Between the publication of *Sissie* and the critique of King, Williams returned to Syracuse in October 1963, living for a month in a neighborhood a few minutes' walk from the 15th Ward. Although Williams had attended closely to Syracuse in *Sissie*, he had been away for a decade. The timing of his return was not random. Slum clearance was ramping up, and a month earlier, Syracuse's CORE chapter had announced its intention to "halt construction at city urban renewal sites." CORE spokesman George Wiley told Mayor William Walsh to expect "direct intervention," and on the first morning of organized protest, more than two hundred people took to the streets. Walsh called this an unacceptable affront to "the duly constituted government of the City of Syracuse." The mayor announced the city would not tolerate any further disruption to the "orderly processes" underway in the 15th Ward, and the police promptly arrested ninety-eight marchers.[79]

FIGURE 31. CORE protestors picket at a 15th Ward demolition site in Syracuse, New York, 1963. A crane and partly demolished building are visible at right. Courtesy of the Onondaga Historical Association.

CORE was undeterred, as were the activists who continued to show up at People's AME Zion Church to organize resistance to slum clearance. Many were community members, and Syracuse University students—carrying on Feiner's legacy—made up a large contingent, too. The church, located just five blocks west of where the marines destroyed the McClouds' home, was an epicenter of radical Syracuse. Much to the consternation of city officials, in August 1963 buses had lined up in front of its doors, taking CORE members and fellow travelers to join the March on Washington for Jobs and Freedom. Now, a month later, the church coordinated Syracusans for direct action nearer at hand. CORE had legal counsel on site, coaching the protestors in arrest avoidance—and what to do if they ended up in jail. CORE had no bail fund in Syracuse. The legal situation, still shadowed by memories of Feiner's arrest and trial, had already drawn national attention. By mid-September, both the New York chapter of the ACLU and the NAACP had established a presence in the 15th Ward, preparing to offer legal aid. Perhaps reassured by such steps, marchers departed the church at regular intervals, heading off to picket demolition sites in the 15th Ward or outside city hall. CORE's leaders encouraged them to chant simply, "Freedom! Freedom!" Newspapers reported that counterprotesters lined the streets, rhyming back, "Beat 'em! Beat 'em!"[80]

This was the backdrop for Williams's visit, and at the end of his month-long stay he published his reflections on what he found in "Portrait of a City: Syracuse, the Old Home Town." Part memoir, part history, part indictment of postwar urban America, the article appeared in 1964. By design, Williams rented an apartment "in the heart of the Italian section," above Stagnitta's Music Shop on North Salina Street. He chose the Italian area because it "was proving to be the most difficult to integrate." His aim, evidently, was to get as uncompromising a view of Syracuse as he could find. Williams found the Stagnittas to be friendly neighbors. But he also encountered recurring police surveillance, with cruisers parked outside his door in the evenings, their occupants staring steadily at him.[81] The experience reminded him of the impression he had formed as a welfare case worker: The 15th Ward felt like a "prison," hemmed in by the white, Republican city establishment in command of a police force whose "corruption flourished openly." Little seemed to have changed by 1963. During his month in the city, Williams saw police stop and search Black 15th Ward residents on three occasions. (And Williams spent much of his month in other parts of the city. Presumably such harassment was a constant feature.)[82]

Imprisoned as the 15th Ward was, however, Williams remembered it as a diverse and friendly place. "Eastern European Jews . . . Polish families, a few Indians, Italians, Irish" had been his neighbors as a child. What held them together was simple: "We were all poor."[83] He recalled a social world with solid routines, parents departing for domestic work or manufacturing jobs in the mornings and returning home for dinner, "their faces ringed with sweat and dust." When children were not in school, elder generations watched over them. "Ours was a community, despite everything else, in which survival of the other fellow or his children meant survival for you," he wrote. When Williams returned in 1963, he still knew the few remaining shopkeepers by name—and they remembered him, too.[84]

As Williams narrated his return, he did not hide his astonishment and despair that the 15th Ward was not long for the world. "Although the city fathers came to the conclusion that The Ward had to go—and it had deteriorated badly—great strength emanated from that polyglot community," he wrote. Why could the city not recognize all the good that had existed in the 15th Ward and invest in it? In the absence of that dream, Williams feared that nothing good could come of demolitions and displacement. Looking around his old neighborhood, he caught sight of the title of his next major novel,

FIGURE 32. John A. Williams, photographed through a broken pane of glass, stands on the porch of a home that would soon be demolished to make way for the Interstate 81 viaduct, 1963 (image published in 1964). Dick Bandy / *The Post-Standard* / Syracuse.com.

reflecting that "nothing so badly demoralizes a person who cries out, 'I AM! I AM!' as being ignored."[85]

To be ignored seemed to be the fate of the 15th Ward's residents and those who joined them in calling futilely for its protection and rejuvenation. This condition—being passed over, in an era of increasingly loud cries for equal voice and opportunity—was at the center of Williams's examination of the 15th Ward, trying to take stock of it in its last days. It had changed: White flight had pulled along many of the Jewish, Italian, and Irish residents into more prosperous neighborhoods outside the city center. At a home marked for I-81 demolition, Williams stopped to stand on the porch and look around him. A photographer for *The Post-Standard* captured him leaning over the railing, eyes cast down, brow furrowed, cigarette smoke pouring from his lips. He wore a crisp white shirt; a shiny watch gleamed at his wrist. Since the last time he lived in Syracuse, Williams wrote that he had "crossed five of the seven seas" and traveled to twenty countries on three continents. "And now I was home again." What struck him was noise: "You hear machines tearing down or building up," he wrote. "Urban renewal and new highways have leveled most of it to the ground. Yellow bulldozers groan back and forth, making sure that every last vestige of the unhappy Negro who, toward the end, alone inhabited The Ward, is now crushed out of sight, out of memory."[86]

*

The promises of urban renewal in the 15th Ward were as false as they were everywhere else. And even apart from its socially destructive effects, the I-81 viaduct was an infrastructural disappointment almost from the moment it opened to traffic. By 1972, just four years after the viaduct's completion, the *Syracuse Herald-Journal* compared driving on it to playing "Russian Roulette." It already had a reputation for frequent and severe accidents due to its narrower-than-normal lane widths, curves, short exit ramps, and poor signage. In their rush to pave, Syracuse officials had pushed the limits of interstate design standards and apparently gotten away with it.[87] After another decade, the city experimented with closing exits, thinking this might reduce viaduct traffic. It did not. In fact it made congestion and collisions worse as interchanges at either end of the viaduct handled even more vehicles. Eventually, two lanes were added, but without substantially widening the viaduct this did little to improve the flow of traffic or reduce accidents.[88] By 1990, the viaduct looked like little more than a tax drain. The traffic congestion and accidents were costly, and serious engineering problems revealed themselves. Of particular concern was the viaduct's surface, which had been constructed in two separate layers. In the harsh Syracuse winters, the city coated the narrow, curvy viaduct in copious amounts of salt in hopes of reducing accidents. Over time, the salt seeped in between the two layers, wearing away the entire structure's interior while the damage appeared relatively invisible from the top.[89]

The viaduct's social damages became apparent just as rapidly as its physical defects. I-81 had put surrounding suburbs—where Walter Carroll had found realtors refusing to show homes to Black families—in reach of the city center for commuters.[90] Syracuse's white population fell 20 percent between 1950 and 1970. After the viaduct was complete, it dropped off another 50 percent between 1970 and 2010. Over the same period, the city's Black population grew tenfold, climbing to 42,000. The 15th Ward's homes and businesses may have been gone, but the population was rising—in a poor and destabilized urban environment. White flight had drained off wealth and tax revenue to surrounding areas, taking business and public services along with it.[91] In the city, the number of high-poverty census tracts rose 150 percent between 2000 and 2013, from twelve to thirty. Out of the nation's hundred largest metropolitan areas, Syracuse showed the highest level of concentrated poverty—a characterization of poverty rates over 30 percent—among Black and Latinx residents.[92] By the twenty-first century, problems like these became as obvious as the viaduct's rusted steel and crumbling concrete.

In 2008, local, state, and federal highway officials formed the Study Advisory Committee (SAC) and began a five-year analysis of the viaduct's physical condition and community impacts. They determined that the viaduct was "severely deteriorated due to age and wear" and near the end of its "intended lifespan." Significant portions of its 1.4-mile length were "structurally deficient" and risked imminent collapse. The SAC found that the viaduct was not only crumbling but "non-standard" by interstate highway design specifications, marked by "poor sight distance, sharp curves, and limited ramp spacing."[93] The viaduct's origins lay in slum clearance, not engineering genius. The SAC did not explain why or how the viaduct had evaded interstate design requirements. But officials reached a conclusion: It was time for the viaduct to go.

The SAC identified three possible paths forward. The viaduct could be rebuilt; it could be replaced with a below-grade highway in a tunnel or trench; or it could be removed entirely. The SAC recommended the third option. This looked to be the most cost-effective. Simply removing the viaduct would cost less than removing *and* rebuilding it, or replacing it with a tunnel or trench. And after its removal, the SAC pointed out, social repair work could follow. The 2013 report was among the first official statements that recognized the long history of destruction that stemmed from the viaduct. The report noted that I-81's route had imposed "barriers" between different Syracuse neighborhoods. Tearing it down would "improve connectivity for the residential areas" and enhance the city's "social equity." I-81 could be rerouted around the city on the existing path of I-481, a beltway in the far less dense city outskirts ready to be widened if necessary to accommodate more traffic.[94]

With crumbling infrastructure came an opportunity to imagine a different future for Syracuse. But progress arrived slowly. It took five years to move from the initial recognition of the viaduct's failures to the 2013 report calling for its removal. It took another five years for clearer plans to develop. By 2018, there were proposals for a "community grid" to replace the viaduct and reconnect the streets and neighborhoods it had severed.[95] Nothing was final yet, though. Highway officials may have recommended viaduct removal, but the state had not yet committed to the plan. In Syracuse, the divisions long structured by the viaduct played out in disagreement over its removal. Political organizing would be necessary.

The *Post-Standard* took the lead in endorsing the community grid plan in 2018, predicting it would knit Syracuse back together and spur "inclusive growth."[96] Soon after, three of the city's most important financial engines—Syracuse University, Upstate Medical University, and SUNY College of Environmental Science and Forestry—followed suit, supporting the plan to remove the viaduct that had long separated their campuses from the 15th Ward (now known as the East Adams neighborhood). The list of endorsements started to grow: St. Joseph's Hospital Health Center, a major employer; the Gifford Foundation, the city's preeminent philanthropy; the Chamber of Commerce.[97] And in the spring of 2019, the state formally committed to supporting the community grid plan, in addition to viaduct removal.[98] Much as federal funding had once helped states fall into line in interstate highway construction, now state funding incentivized Syracuse to support the SAC's decade-old recommendations.

It was one thing to have glossy new plans for change and growth. As the long history of the interstate highway system showed, turning such plans into concrete realities that actually worked for communities was another matter altogether—and not one with a good track record. It was remarkable that the state and city had lined up behind the community grid plan. But the real work of constructing community equity had only just begun. "Racial and environmental justice is not going to happen just because you remov[e] the viaduct," warned Lanessa Owens-Chaplin, director of the Environmental Justice Project at the NYCLU.[99] The process of removal would amount to a vast and imposing demolition process, undertaken in close proximity to Syracuse's most vulnerable residents. And real guardrails would have to be erected to protect the community from new patterns of gentrification-borne displacement when the community grid started to take shape.

*

The new plans in Syracuse were pathbreaking, but the NYCLU recognized there was a likelihood past mistakes would simply be repeated. The NYCLU's Central New York chapter was formed in 1963, in response to police harassment of the viaduct protestors as the project got underway.[100] Now the viaduct was on the way out, and the NYCLU had new work to do ensuring that the community had a say in the equitable redevelopment of the area. In 2020, the organization released a landmark report that took stock of the problems stemming from the viaduct and identified challenges going forward. The goals it expressed were a rejoinder to the entire paradigm that had ruled interstate highway development since its inception. The NYCLU's engagement in Syracuse aimed "first, to share information about the I-81 project with community members who are often left out of public conversations. Second, to listen to their concerns and questions and relay those comments to NYSDOT [New York State Department of Transportation]. And finally, to build trust with community members to help achieve their hopes for the neighborhood."[101] The three points confronted the absence of community participation that had marred the interstate highway project nationwide.

The NYCLU spent two years hosting "community conversations." Going door to door, staff asked questions about life under and around the viaduct and conducted lengthy interviews with residents. The organization developed a resource-rich hub through which community members could share knowledge, collaborate with one another and with partners like the NYCLU, and devise institutional channels to amplify their voice to state and federal highway authorities.[102] The NYCLU also partnered with the Onondaga Historical Association to create an exhibition on the 15th Ward's history. The exhibit traveled to New York City and Albany in an awareness-raising effort. Its purpose was to explain "I-81's legacy of inequality" and reveal the uneven patterns of poverty and prosperity the viaduct shaped in Syracuse.[103]

When the NYCLU released its report, it found the viaduct responsible for three categories of harm: a racial wealth gap, segregated housing, and health impairment. At the time of the report's publication, 40 percent of Syracuse's Black population lived below the poverty line, compared to 11 percent of white Syracusans; about 30 percent of Black residents owned homes in the metropolitan area, compared to over 70 percent of white people. Syracuse continued to exhibit the highest level of Black and Latinx concentrated poverty of any of the nation's hundred largest metropolitan areas. That racial wealth gap was closely connected to long-term housing segregation. The 15th Ward's history was defined by redlining and other forms of housing market discrimination that resulted in underinvestment and substandard conditions. Property values were never high, and federal officials routinely paid less than market value during right-of-way acquisition during the viaduct's construction. On top of that, many community members were renters and received no compensation at all. Once constructed, the viaduct's presence lowered remaining property values even further. Displaced residents generally had limited relocation resources and found their way to even lower-value properties farther from the city's center. All of this amounted to a "generational loss of wealth accumulation."[104]

While the viaduct's presence ensured that wealth could not compound over time in the 15th Ward, the same presence meant that health problems would. When property values around the viaduct declined, the city took advantage of the low-cost, politically disenfranchised environment to site a sewage treatment facility and steam plant in the area, which further depressed property values. These became factors in the area's elevated rates of lead poisoning and asthma—although the worst culprit was emissions from the highway itself: Decade after decade, heavy toxins from the viaduct's traffic landed on the community below. The pollution was not just a matter of bodily health, either. People in poor health are likely to miss work, making it harder to hold down steady jobs and climb the economic ladder. Education suffers, too. The surrounding school district's asthma hospitalization rate was among the highest in the state by the time the NYCLU released its findings. There are empirical links between poor air quality and academic performance. Since 2011, the EPA had issued pollution reduction guidance to schools within six hundred feet of major roadways. The viaduct passed less than two hundred feet from Dr. King Elementary School on McBride Street—seven blocks from where Irving Feiner was once arrested. The irony of the school's name, in light of John A. Williams's perspective on Martin Luther King Jr., is poignant. When the NYCLU issued its report, the school had not been able to secure funds to implement the EPA's guidance.[105]

The NYCLU made a powerful case for viaduct removal, connecting racist past choices to present-day inequality. The NYCLU also raised serious concerns about how to buffer the community from the long job of viaduct demolition, a process that would fill the air with noise, dust, and toxins throughout the years-long removal and construction process. And once that phase came to an end, the NYCLU underscored, there would be a need for safeguards against gentrification, likely involving a land trust that gave community residents a measure of direct involvement in land use decisions. Despite these challenges, the NYCLU saw hope in the community grid project.[106]

*

The NYCLU's work underscored the scope of the 15th Ward's historical disenfranchisement. The 2020 report illuminated a range of historically embedded issues to be redressed and identified perils to be avoided as the community grid plan took shape. The organization's findings also produced a preemptive bulwark against pushback that the community grid plan encountered from factions that preferred to keep the status quo in place. After the state signed on to the community grid plan, opposition quickly turned litigious. Suburban residents argued that the community grid would lengthen their commutes into the city center. Their complaint bore out a central premise of the interstate highway system: that suburbs could become effective parts of national political economy thanks to highways that simultaneously blockaded some people while easing the mobility of others.[107] Some business owners reported that they had "staked their businesses near the off-ramps" along the I-81 corridor; relocating I-81 would dash their business plans.[108]

Suburban and business interests banded together in 2022 to form a group called "Renew 81 for All," headed by a former Syracuse police chief. In place of the community grid, they offered a vague proposal for a "sky bridge" rising some fifty feet taller than the existing viaduct, carrying traffic through the city on a massive new structure they suggested naming the "Harriet Tubman Memorial Freedom Bridge."[109] Renew 81 for All's main tactic was to weaponize the National Environmental Protection Act. The group claimed in state court that environmental review for the viaduct removal and community grid plans had failed to meet requirements, and that the state therefore had to begin all over again in its planning. Considering that fifteen years had passed since the state first started studying the viaduct's deficiencies, the stakes were high. The NYCLU threw its weight into the case as well, submitting an amicus brief that reiterated many of the points from its landmark 2020 report.[110]

Renew 81 for All succeeded in convincing Judge Gerard Neri of the New York State Supreme Court (the state-level trial court in New York) to require some additional environmental review. Neri found no issue, however, with the review that had already been conducted, meaning the upshot was only a marginal victory for Renew 81 for All. Not wishing to take on additional review, the state appealed the case, and five appellate court judges unanimously struck down Neri's decision in 2024.[111] No further environmental review would be required. As a last resort, Renew 81 for All filed a federal case on similar grounds but quickly dropped it, without explanation. In the end, Renew 81 for All succeeded only in delaying the project—proving yet again the difficulty of defeating highway development, even if now the substantive and normative aims had changed compared to the days of decisions like *Nashville I-40 Steering Committee v. Ellington*, in 1967.[112]

Other opposition was less litigious but more indicative of enduring problems of inclusion. Joe Driscoll, a former Syracuse Common Council representative who served as project director for the viaduct removal project, encountered grave concern in the 15th Ward itself. "A lot of the African American community in that corridor, are deeply distrustful of government in all its forms, particularly white folks," he noted. Gentrification was the key concern. Rumors spread that the land freed up by viaduct removal was already contracted for sale to Syracuse University for student housing. Charles Garland, an Onondaga County legislator representing a majority Black district, claimed that the community grid project's only aim was "taking down the viaduct so University Hill can be extended into downtown and they can appropriate the surrounding properties."[113] It was unclear whether this rumor sprang from real fears and suspicions or had a malicious origin among those opposed to tearing down the viaduct. But either way, it spoke to generations-old tensions and divisions. Driscoll summed up the prevailing sentiment among suspicious 15th Ward residents as "you've just come here to chase us out of our homes and knock them down." The city did little to assuage such concerns.

In 2009 and 2010, in the early days of the SAC's work imagining alternatives to the viaduct, the group conducted twenty-three focus group meetings throughout Syracuse in an effort to learn about residents' concerns. Later, the SAC held workshops and public meetings.[114] The input had no binding impact on planning. But the quantity and variety of participation the SAC sought was striking. More than a decade later, after the community grid plan was official, opportunities for participation dried up. In 2024, when construction was poised to commence, the city offered only five ninety-minute "open houses" in which residents could learn about the project.[115] Officials were not even in charge. They handed the process over to a consulting firm, which promptly renamed the open houses "charrettes." The term has origins in participatory conflict resolution processes. But here, participation amounted to the consultants inviting community residents to write their ideas or concerns on sticky notes and paste them onto relevant areas of a large map of Syracuse.[116]

One incredulous activist dramatized the perspective the consultants, at the city's behest, appeared to take: "They [the community] spoke, we listened, then we did our own thing."[117] The richer involvement of community residents over a decade earlier had helped shape the broad contours of the community grid plan. But the crucible moment for real citizen involvement was now, as heavy machinery powered up and real plans for what would come next took shape. It was clear by this point that the NYCLU had enabled more meaningful citizen participation through its two-year-long process of community meetings and interviews. The community knowledge that the NYCLU had surfaced through that process clarified the potential downsides that the community grid plan could bring—and now there was little formal means by which the community could confront and avert the looming perils.

There have been flashes of hope in Syracuse. The community grid plan initially located a busy new traffic roundabout less than 150 feet from Dr. King Elementary School, which for decades suffered alongside the viaduct. Activists quickly protested the choice. The NYCLU once again stepped in to advocate and succeeded in convincing officials to move the roundabout farther away.[118] The incident showed the alacrity with which the community could identify emerging threats to the new era of equity promised by the community grid and organize against them. But it also showed just how tenuous that equity was—and how little structural capacity existed for citizens to lodge their concerns directly within the halls of government. Once again, the NYCLU played a role that might have been filled by a democratically mandated government apparatus more closely and authentically engaged with citizens.

Owens-Chaplin, the NYCLU lawyer, emphasized that the area's residents had been "historically ignored or left out of government decision-making processes" and warned that "past mistakes" were on the verge of being repeated, despite her belief in the community grid's potential. In the short term, the NYCLU and the community had raised the alarm about lead exposure that would come from tearing down a highway soaked in decades of gasoline—and that was to say nothing of more basic concerns with noise, dust, and the inconvenience of heavy machinery working just feet away from peoples' homes for years to come. City officials issued a statement claiming that "the community will not be impacted by construction." It was another example of government by promise, not participation.[119]

*

The community grid is a step toward a new paradigm of infrastructural justice. It has far more upside than the viaduct ever did and may well produce a vibrant, mixed-income, mixed-use reincarnation of the 15th Ward. But whether it fulfills that vision hinges on solving the inequities of private development, gentrification, and a host of other problems that might not even be foreseen.[120] "If it's done right," Owens-Chaplin averred, "what happens in [Syracuse] will really be a model and the beacon of what can be done for the rest of the nation."[121] But one thing looked sure to her: The community could not simply trust the city to retain control of the land and "do the right thing."[122]

One of the consultants overseeing the community grid "charrettes" told Syracuse's *Post-Standard* that the aim was for the city to become "the model for every other city in the nation that's taking down a highway"—a growing list.[123] The Congress for the New Urbanism reported ten other cities, from Oakland to Tulsa to Albany, following in Syracuse's footsteps.[124] But exactly what kind of model Syracuse provided for interstate removal projects remained to be seen. The community grid plan did not resolve deep problems with the community's role in city government—much less state or federal decision-making.

If the community grid was a partial victory, its shortcomings clarify the governmental problems yet to be solved involving democracy, participation, and equity in the work of infrastructural development—at metropolitan and national scales alike. The best-case scenario in Syracuse stands on shaky ground, destabilized by fault lines that stretch far into the past and continue to shape the landscape today. Like countless other American communities, the 15th Ward never had much of a chance as the mid-twentieth-century infrastructure boom took hold. "It is imperative that the history of redlining, urban renewal, and displacement in Syracuse that has disproportionately hurt Black neighborhoods does not repeat itself," wrote the NYCLU in 2020.[125] If Syracuse actually intended to redress the viaduct's decades of harm, the most important steps would amount to fair housing guarantees and community control over the ensuing development process once the community grid was in place.[126] In other words, citizens need meaningful input in government.

Syracuse's removal of the I-81 viaduct is a momentous achievement in the fight for infrastructural justice, and the community grid plan is a topical remedy for the area's spatial and social ills. But it is not a solution to the foundational challenge of democratizing government and developing participatory processes better equipped to avoid community harm and build community equity in the first place. That work remains to be done: The question, in the aftermath of Syracuse's pathbreaking efforts, is what kinds of legal and governmental mechanisms can be designed to bring communities into the work of state building.

CONCLUSION

End Road Work

In 1938, James Landis—New Deal administrator, legal scholar, and progenitor of the modern American administrative state—wrote that when it came to "controlling human affairs," it was easier done "in the comparative quiet of a conference room" than "amid the turmoil of a legislative chamber"—to say nothing of a democratized institution filled with a welter of citizen voices.[1] Landis's famous statement reflected the era's reckoning with the Great Depression, in which administrators stemmed the tides of crisis and made abstract ideals of high modernism concrete. Expert authority, apparently, could manage political economy for the benefit of society as a whole. The "administrative ordering" of the national geography and its economic life unfurled with force in the form of the interstate highway system.[2]

But problems quickly became evident. Congress, much less the public, had little say in how the interstate highway project proceeded. Plans called for bulldozing homes and businesses and paving over communities to make way for the new highways. A few months after that dual process of destruction and construction began, the American Road Builders Association (ARBA) encouraged workers not to worry about the havoc they were wreaking. "A few must suffer so that many may benefit," the group averred.[3] But more than a few suffered: The interstate highways displaced over one million Americans and disrupted the lives and livelihoods of countless others. And the burdens were not shared equally.[4]

As the interstate system took shape, it embodied a racist and undemocratic state-building regime. Activists pushed back as best they could, hoping the era's heroic civil rights victories would establish a legal framework that protected their communities from targeted harm. But cases like *Nashville I-40 Steering Committee v. Ellington* revealed foundational problems with democracy, participation, and equity. Today, in *Ellington*'s long aftermath, it has become clear that current legal frameworks and administrative agencies are not up to the task of solving those problems and protecting communities from infrastructural injustice. Civil rights law and environmental law, which look on the surface like promising tools, have failed to provide community protection. Existing institutions do not engage communities on participatory terms that might be sufficient to democratize decision-making and achieve community equity. But there are aspirational solutions on the horizon. Examining what has failed marks out a clear blueprint for an alternate administrative and infrastructural future capable of distributing the burdens and benefits of infrastructure projects more evenly through society.

*

The problems start and end with community involvement in the work of infrastructural development and state building. When *Ellington*'s plaintiffs challenged racist choices and disparate impacts in 1967, they revealed the significance of the interstate highway project's exclusion of meaningful community participation. Extending from the problem of participation, *Ellington* indicated civil rights law's limitations in dealing with infrastructural harms. Concepts like equal protection and due process—central to the pathbreaking Civil Rights Act of 1964—could be foiled.[5] Another problem was civil rights law's emphasis on individual claims, usually relating to property or health. The civil rights framework might ensure that someone received compensation when officials took their house to make way for highway construction. But individual claims are not an effective surrogate for community protection. For communities, the acknowledgment of individual harm has not been sufficient to block development—or gain a voice in how development proceeds.[6]

A year after *Ellington* indicated the sharp limits of civil-rights-based community involvement and protection, the National Environmental Policy Act of 1969 (NEPA) injected a wide range of concerns into infrastructure projects. It was a well-intentioned piece of policymaking, aimed initially at measuring the ecological costs of development projects.[7] But it did not resolve the problems *Ellington* revealed relating to participation and equity in the work of infrastructural state building.

NEPA affects infrastructural development through its core requirement that agencies produce "environmental impact statements" (EIS) whenever federal activity will impact "the human environment."[8] This requirement is construed broadly: Environmental impact includes social, economic, and cultural changes, among other categories.[9] However, the statute only requires agencies to complete the EIS process and identify the potential effects of the federal activity in question. The Supreme Court has clarified that NEPA merely "ensures that the agency, in reaching its decision, will have available, and will carefully consider, detailed information concerning significant environmental impacts." Reminiscent of the talk of "wisdom" versus "law" in the *Ellington* decision, the Supreme Court averred that NEPA prohibits "uninformed—rather than unwise—agency action."[10] NEPA does not prevent agencies from taking their desired action, even if a problem is identified. The EIS has enormous reach, tabulating a vast range of legitimate issues that may spring from federal development, but carries little to no substantive power.[11]

The NEPA framework, at the federal level and through state-level counterpart laws, can also be weaponized. This was the case after activists and officials in Syracuse opened a new horizon of infrastructural justice. Plans to tear down the Interstate 81 viaduct and replace it with the "community grid" that repaired local streets and neighborhoods long since destroyed by interstate construction became an instant model for redressing past infrastructural harm. But some businesses and suburban commuters preferred to keep I-81 in place. A group of plaintiffs tried to block the community grid project by filing a lawsuit claiming environmental review requirements had not been followed.[12] Their aims were transparent. As the case unfolded, Syracuse's mayor—who supported the community grid—reminded the public that the plaintiffs' only tactic was "delay, delay, delay."[13] The Syracuse community grid plan prevailed, but in many other cases NEPA's weaponization has succeeded in thwarting projects for reasons that have nothing to do with ensuring the health of the "human environment" that the statute claims to protect. While a project is stalled, funding can dry up, political power can shift, or other challenges can take hold.[14]

Of course, environmental claims-making has its upsides, most of all when it is based on genuine environmental concerns. And whether or not ecological and social degradation is the real concern, if the goal is to halt highway construction, erecting procedural roadblocks can be beneficial. But the EIS is not an effective mechanism for protecting or producing community equity. Even setting aside that substantive goal, the need to weaponize procedure is hardly indicative of a well-functioning legal and governmental apparatus charting the course of infrastructural development. In Syracuse, for example, a different approach—one amenable to community-level analysis and citizen participation—might have forced those who wanted to retain the viaduct to make an affirmative argument in support of their vision of the city's future. And, in response, the defense might have mounted an equally affirmative legal argument in favor of the community grid, rather than simply arguing that environmental review procedures had been followed dutifully. The substance of competing goals was present in the litigation only by inference.

NEPA's structure means officials can ignore EIS findings, and plaintiffs can waste time and resources by filing lawsuits making claims that procedures have not been followed. Plaintiffs, in such cases, do not even need to explain what kinds of outcomes they find objectionable—or what kinds of alternatives they prefer. They can simply derail projects, sometimes permanently. The EIS framework, measuring impacts but requiring no action on behalf of the knowledge generated, does not offer a means of critiquing an existing infrastructure regime, nor of constructing a new one.

*

If infrastructural justice is the aspiration, the reasons that the existing legal framework cannot achieve it are clear. Public hearings compel no action on the part of officials, who also have ample opportunities to manipulate the proceedings.[15] Atomized civil rights claims are poorly equipped to counter community-level problems. The civil rights framework has largely failed to provide collective protection against infrastructural development schemes and is an ineffective tool for balancing burdens and benefits and achieving equity. NEPA's EIS requirement does not provide an affirmative and enforceable path to equity, either. And it can be turned *against* that ambition when plaintiffs seek to undermine infrastructure that aims to redress existing patterns of discrimination.

What kind of legal framework would be better equipped to balance infrastructure's burdens and benefits? Legal scholar and ACLU President Deborah Archer has been at the vanguard of exploring answers to this question. The goal is to "maximize the opportunities for communities to advocate and make decisions together" in pursuit of infrastructural justice.[16] Archer calls for a "racial equity impact study" (REIS) requirement similar to the EIS approach. But where the EIS is procedural and unenforceable, Archer's REIS is modeled on state-level laws that emphasize substantive issues and require action in response to findings. California law, for example, explicitly directs state agencies to mitigate or avoid harms uncovered in state-level environmental impact statements. There are substate models, too. In Seattle, officials must evaluate "equity and social justice" and mitigate disparities in development projects, and municipalities in Oregon, Minnesota, and Wisconsin have followed with similar laws.[17]

Scaled up into a national framework, the REIS would aim to "systematically analyze how racial and ethnic groups will be affected by a proposed action, policy, or practice." By identifying, evaluating, and communicating the consequences of an infrastructure project, the REIS would enable a broad understanding of community harms and indicate a "range of interventions available to resolve them equitably."[18] Significantly, the REIS framework recognizes infrastructure as a meeting point of past choices, present realities, and future problems, eschewing narrow snapshots and surface-level scrutiny. "Racial impact studies," Archer writes, "not only identify potential racial disparities, but can also identify the often-invisible historical influences, systemic inequalities, structures, and institutions that interact to create those disparities."[19] Synthesizing past and present-day demographic data, the REIS would assemble a comprehensive portrait of impacts on everything from housing to education to health to economic opportunity.[20]

This methodology would generate rich, community-level data, which would enable rigorous, substantive legal processes regarding the location of a new infrastructure project or the abatement of an old one. In other words, it would start to transcend the procedural limitations of EIS-based litigation. But if the REIS can compensate for the frailty of civil rights law and NEPA in dealing with infrastructural harm, success will depend on two challenging factors: the involvement of citizens in the REIS process and the enforceability of the findings they reach in collaboration with officials.

To effectively identify and counter community harms, the REIS process will have to be a community-based one, in which citizens articulate and officials respond to harms—harms that may be difficult to see or quantify for those who do not spend their days and nights in the community. If communities become meaningful stakeholders in the REIS process and contribute to the data it generates, policymakers can make equitable decisions—if they are compelled to act on that information. The REIS framework is only useful if it actually alters priorities and changes decision-making. Otherwise, it will fall to the same ineffectual (and possibly counterproductive) fate as the EIS. As Archer has underscored, community members "must not only be heard, but have a seat at the table and be involved in the substantive conversations and decisions that will shape the direction of the project."[21] The surest way to accomplish this is to embed the REIS and its participatory processes within a federal agency tasked with synthesizing grassroots and expert knowledge and upholding community equity.

*

The administrative structure that brought the interstate highway system into existence took every opportunity to insulate itself from political pressures, including citizen input. But there is a long, if frequently suppressed, American tradition of democratizing agencies and bringing citizens into the work of administration and state-making. During World War II, the Office of Price Administration (OPA) relied on citizen volunteers to control prices and ration supplies. The Economic Opportunity Act of 1964—the cornerstone of President Lyndon B. Johnson's vision for the Great Society—established over 1,500

"Community Action Boards." Requiring "maximum feasible participation" of the poor, the aim was to make citizens central participants in directing federal economic policy toward the ambitious goal of eliminating poverty.[22]

A more recent example is the Consumer Financial Protection Bureau (CFPB), formed in the aftermath of the 2008 financial crisis. The agency's guiding legislation positioned it to represent consumer interests against "unfair, deceptive, or abusive financial practices."[23] This required an administrative blueprint capable of blending together a substantive mission, a commitment to grassroots engagement, and an insulated funding stream channeled through the Federal Reserve Bank that aimed to protect the agency from corporate capture and political meddling. The last feature is a fraught characteristic reminiscent of the Highway Trust Fund, among other examples. The interstate highway system's history shows that such fiscal insulation can exacerbate the administrative state's undemocratic possibilities. So, the merits of the CFPB's model hinge on the agency's faithful representation of consumers within the financial regulatory system. The CFPB's creators sought to ensure its mission fidelity—and thus the upside of its insulated structure—with three pillars of institutional design: centralized authority, clear lines of accountability and responsibility, and—crucially—citizen involvement. That approach made the agency's activities a resounding success over the CFPB's first decade and a half of existence.[24]

As this book went to press, the CFPB was clinging to life, besieged by Supreme Court efforts to undermine the administrative state and by the second Trump administration's deregulatory agenda. But the CFPB has already shown that it is possible to craft a federal regulatory apparatus "oriented toward participation and engagement," as legal scholars Sabeel Rahman and Hollie Russon Gilman put it, and capable of enhancing the nation's democratic capacity.[25] The CFPB remains the best available model for expanding citizen-centered administrative governance to other areas of government. One practice demonstrated by the CFPB, with obvious applicability to a REIS-based regime of infrastructural development, is the use of proactive, rather than reactive, public hearings. On this front, the CFPB devised a system of "field hearings" around the country. The purpose was not to solicit communities' reactions to official plans, but rather to collect information about the priorities and policies that communities wanted to see enacted. This simple inversion of the standard public hearing process centered citizen knowledge and input. This helped make the agency into a "proxy advocate" through which citizens could file complaints and redress problems.[26] Working in concert with this approach, the CFPB also built a publicly searchable database of complaints. This flow of information, from the grassroots into the agency and back out to the public, harmonized citizen knowledge with agency enforcement capabilities to challenge exploitative practices.[27]

The relevance of the CFPB model is evident. Imagine the creation of a "Community Equity Protection Bureau." This new federal agency would be designed to engage communities in collaborative knowledge formation and democratic decision-making. Its mission would be to uphold community equity. It could build on the CFPB's successful approaches for synthesizing grassroots and technocratic knowledge into policy capable of realizing a substantive social vision. With the REIS process at its core, the Community Equity Protection Bureau could employ proactive hearings and public database techniques to provide a bulwark against infrastructural harm and a path toward infrastructural justice.

Of course, the Community Equity Protection Bureau would differ in key ways. The CFPB was enacted to regulate private entities through direct legal action. The Community Equity Protection Bureau might find itself confronting private actors and subnational authorities in a similar manner. But it would also be arrayed against other federal agencies. In this situation, it would likely not conduct enforcement directly. There is precedent for federal agencies to sue one another, but the scenario is complex and significant legal work remains to be done to bring predictability and utility to that course of action.[28]

But even if the Community Equity Protection Bureau served as a clearinghouse at the federal level for tabulating community harm and advancing equitable alternatives, it would play a powerful role in legal action. Proactive hearings could generate community-based policy proposals, and the agency could provide a structure for aggregating complaints and identifying patterns of discrimination and disparity that find purchase in community-based litigation—following a model described by legal scholars Matthew Cortland and Karen Tani.[29] The Community Equity Protection Bureau help would assemble precisely the kinds of community-level claims that have failed to cohere in the context of civil rights litigation, while aiming at substantive goals that are inherently absent from NEPA lawsuits. From this institutional framework, a new field of litigation would emerge around community equity.

A promising area in which this litigation might begin is public nuisance law. Legal scholars have begun to explore how this body of law can be applied to community protection in the face of infrastructural harm. In brief, public nuisance claims are pursued successfully when there is "significant interference with the ability of community members to access, use and enjoy public spaces and resources that they are entitled to access, use, or enjoy."[30] Public nuisance recognizes the possibility of community rights, and thus community harms—marking a crucial divergence from civil rights law's emphasis on individual property and well-being. Ironically, public nuisance law first came into use through claims rising from interference with communities' ability to use public roadways. Can that example be flipped on its head? If a highway development project can be shown to abrogate the ability of a community's collective members to use its spaces and resources, perhaps a public nuisance case can be made.[31]

The nation's moves toward infrastructural justice range from the tentative to the aspirational. Local and state-level REIS frameworks exist; they can be scaled up into federal law. Public nuisance law has a long history of successful litigation; it is ready for communities to deploy in battles against infrastructural harm. The steepest challenges surround democratic administrative governance. The OPA's volunteer structure and the Great Society's Community Action Boards were eventually subsumed by a combination of conservative politics, private power, and elite administrative control.[32] The second Trump administration made the CFPB's destruction one of its first aims in 2025, pulling every lever it could find to unwind the agency's legal and financial framework.[33] This assault was only the latest example of American antipathy toward institutions that democratize power and decision-making—especially when such institutions are intended to check profit seekers and adherents to the gospel of growth that animated the interstate highway system from the start.

*

Democratized administrative institutions require careful design. James Landis may have been right that it is easier to make decisions and govern society from the calm of a conference room than amid the frenzy of representational, much less participatory, democracy. The goal is not to open every single decision to unbounded debate and inexpert determinations, but to revise a governing paradigm that reduces democracy to a combination of public-opinion surveying and voting. These are important building blocks, influencing and incentivizing official decision-making. But more robust forms of input could be created, especially in systems where significant matters of government are delegated to unelected representatives, who for better and for worse enjoy a layer of removal from formal and grassroots political pressures.[34]

That insulation is often crucial to avoiding cronyism and capture. Good administration requires a degree of discretion and the ability to operate free from patronage or fear of retribution. Administrators are, after all, experts in the craft of government, and they should be empowered to do their jobs skillfully and dutifully. But the history of the interstate highway system also shows the damage that occurs when communities are locked out of decision-making processes of which they will be the prime subjects. It is time to correct the course of government development in this regard, to make citizens more than "clients, voters, or customers," as Rahman and Gilman put it. They should be empowered as problem-solvers and decision-makers in collaboration with government experts.[35]

At first glance, the timing is not auspicious to establish an agency like the Community Equity Protection Bureau. The administrative apparatus responsible for the interstate highway system—the BPR—was emblematic of the "era of deference" in which judges submitted readily to the will of the administrative state in the work of managing society.[36] That era endured for decades. There were some revisions—*Overton Park*, for example, was an important episode in which the judiciary declined to give administrators carte blanche. Still, the overall paradigm changed little. In 1984, the Supreme Court ruled in *Chevron U.S.A., Inc. v. Natural Resources Defense Council, Inc.* that courts should defer to agencies' "reasonable" interpretation of statutes guiding their institutions and actions.[37] "Chevron deference," as it became known, shaped some 18,000 cases over the next forty years.[38] Then, in 2024, it came crashing down, and with it, decades of lawmaking that made administrative governance possible. In the case of *Loper Bright Enterprises v. Raimondo*, the Supreme Court's six-justice conservative wing—animated by right-wing claims of administrative authoritarianism surrounding Chevron deference—ruled that courts, not administrators, are to interpret statutory guidance of administrative activity, creating "judicial control over agency action."[39]

Elevating judges' priorities and interpretations is hardly a recipe for democratizing the administrative state. Expanding participatory government of the nation's people and affairs was not the motivation for the conservative Supreme Court justices who overturned *Chevron*. The new framework might help to curtail some instances of administrative overreach, but only incidentally and not by democratic means. *Loper Bright* appears to give judges broad latitude to undermine agencies' prevention of exploitation and their protection of vulnerable communities—all the while curtailing the ability of citizens to have a say in such government.[40] Meanwhile, the judiciary, in the elevated position it has given itself, has neither the expertise nor the capacity to effectively manage a nation's worth of administrative governance.[41] It is not yet apparent how the *Loper Bright* decision will play out in full, as it empowers judges to pick winners and losers among agencies and throughout the citizenry. But it is clear that the new paradigm produces a profound shift in political and economic power, making it easier for anti-democratic factions to move federal governance away from normative goals like infrastructural justice and community equity and further reduce the administrative state's democratic capacity.

The history of the interstate highway system reveals deep problems with democracy, participation, and equity. Those problems were foundational to the iteration of the administrative state that rose to power alongside visions of the interstate highway system and then governed for some ninety years. The solution to those problems is not to dismantle the administrative state in its entirety, destroying its best possibilities along with its worst tendencies. But that is a possibility, given the present situation. The challenge now is to fight back against *Loper Bright* with the aim of rebuilding far more than the previous administrative status quo. In Syracuse, the oppressive and dominating Interstate 81 viaduct had to crumble in disrepair before it became possible to build alternate infrastructure capable of supporting an equitable economic, social, and political landscape. With the administrative state in disarray following the *Loper Bright* decision, an analogous opportunity has arrived. The case's outcome prompts fundamental and open-ended questions: What does democratic administrative governance look like? What design features are necessary to balance the burdens and benefits of infrastructural state building throughout society? What kinds of new agencies do we need? These questions should guide the work of building truly democratic government.

For democracy's proponents, the changes and uncertainties brought by *Loper Bright* have removed the need to play defense. There is nothing left to do but go on the offensive and imagine a new administrative and infrastructural future—just as the imagineers of the interstate highway system once did, when they devised new powers and sculpted a new political economy across the national landscape. The task—a mighty one, to be sure—is to invert that history. A new administrative state can be built on principles of institutionalized participation, enabling democratic legal frameworks that construct an egalitarian political economy. From such a foundation, communities may yet realize the full promises of American democracy and shape the arc of their own futures on equal terms.

Acknowledgments

What a long road it has been. As this journey of research and writing comes to an end and *Concrete Leviathan* makes its arrival, it is only fitting to start with my thanks to the people directly responsible for getting this book across the finish line. My gratitude begins with Tim Mennel, my editor at the University of Chicago Press. Thanks, Tim, for believing in this project from a very early stage and guiding me through its completion. Authors can usually only dream of the editorial wisdom and rigor you brought to the process. Our discussions—ranging from the politics of mid-century administrators to comma placement—meant everything. I hope we will get to work together again. At the press, Andrea Blatz masterfully managed the day-to-day of wrapping up the manuscript, and Stephen Twilley's expert copyediting made all the difference.

A book like this cannot exist without the aid of archivists and librarians. I am indebted to the staff at Princeton's Firestone Library, particularly Steven Knowlton. At the Hagley Museum and Library, my thanks to Lucas Clawson. In Kansas City, Ben Gross made my time at the Linda Hall Library enormously productive. My research in Syracuse would not have been possible without advice and assistance from Steve Billmyer, Jonathan Croyle, and Marie Morelli at *The Post-Standard*. The same goes for Maria Lore and Robert Searing at the Onondaga Historical Association, and for Dan Smith at the Onondaga County Public Library. A great deal of my research took place at the National Archives facility in College Park, MD—a vast, impersonal, and peerless institution whose existence gives me faith in bureaucracy. The nature of the place and the specifics of my research meant that I dropped my record requests on a tall pile of other queries from fellow researchers. I have no idea who collected the slips, pulled my materials, and wheeled them out on a cart for me to collect, every hour on the hour. But they made this book possible. And as an edifice of democratic, transparent government, the National

Archives is among the most undercelebrated of American institutions—let us make sure it can continue to do its good work.

A book like this also cannot exist without real interlocution and collaboration. I was fortunate to present research twice at Princeton University's Modern America Workshop. Thanks to the stellar feedback I received, the drafts I shared soon became book chapters. Participation in several conferences refined important aspects of this book, as well. At the American Society for Legal History conference, a panel on race and mobility with Anne Kerth, Justin Randolph, and Kwelina Thompson opened new historical and theoretical terrain for me. At the Society for the History of Technology conference, a panel on postmodern infrastructure with Patrick Chung, Thomas Heller, Zachary Schrag, and Varsha Venkatasubramanian helped me refine key normative arguments. At the American Historical Association conference, joining Destin Jenkins and Cotten Seiler on the panel "Infrastructure, Inequality, and the Politics of Provision" was particularly crucial to the entire final phase of my research and writing process.

Two brief teaching stints also played roles in *Concrete Leviathan*'s completion. At Cornell University, thinking through the vast history of American capitalism with the two-hundred-plus undergraduates who signed up for my lecture course on the subject was a true highlight. My thanks to all of them—and to my terrific teaching assistants, Jeremy Goodwin, Sang Kyung Lee, Michael Moynihan, and Cameron Tardif. My sincere thanks to Lawrence Glickman for bringing me to Ithaca, as well as to Ed Baptist, Ray Craib, Aaron Sachs, and Eric Tagliacozzo. All told, my time on the Hill sharpened several of this book's economic insights and offered me important proximity to my research in Syracuse. From Ithaca, I went on to teach for two years at the New Jersey Institute of Technology. It was a formative stretch for *Concrete Leviathan*. I am so grateful to the nearly five hundred students (yes, I had a very heavy teaching load those two years!) who joined my lectures and seminars on legal history and infrastructure studies. Those many hours of classroom discussions refined every aspect of this book—thanks, especially, to Gabby Zalot for continued conversations about infrastructural inequality. And I am appreciative of Rosanna Dent, Carla Guerriero, Scott Kent, Alison Lefkovitz, Neil Maher, and Stephen Pemberton for welcoming me into the NJIT community.

Concrete Leviathan is the result of nearly a decade of researching, collaborating, teaching, and writing. Throughout, an incredible community of mentors, colleagues, and friends sustained me and supported the book's development. During my time at Princeton University, Margot Canaday offered early and pivotal affirmation that there is indeed much to learn about

American democracy and state building via the study of infrastructure. Margot, it is hard for me to imagine a more skilled historian and mentor than you. You taught me how to ask the right questions—whether of teachers, archives, students, or myself—and how to write history that is attentive to the past and present in equal measures. Thank you for making this book—and all my future work—possible. Princeton was a perfect place to begin my research, and a constellation of faculty and graduate students played integral roles in launching this project—thank you, Ben Bernard, Michael Blaakman, Mike Glass, Joe Glynias, Alison Isenberg, Rob Konkel, Kevin Kruse, Jonathan Levy, Matt McDonald, Jiya Pandya, Marni Sandweiss, Emily Thompson, Peter Wirzbicki, Kyla Morgan Young, and Julian Zelizer.

Beyond Princeton, at key points in the writing process important conversations with Tim Barker, Chris Hughes, Sophia Lee, Mike Perry, and Noah Rosenblum helped me clarify arguments and policy interventions. And during the final stages of bringing *Concrete Leviathan* into existence, I had the pleasure to be working alongside Zohra Ahmed, Corinne Blalock, Matt Dimick, Hiba Hafiz, Luke Herrine, Amy Kapczynski, Luke Norris, Lenore Palladino, Sabeel Rahman, Allison Tait, and Karen Tani—among a longer list of brilliant contributors and fellow travelers—as we built the Association of Law and Political Economy. I could not wish for better colleagues and collaborators. Our work and ongoing discussions have strengthened every facet of this book. In short time, ALPE has gathered a remarkable group of scholars and practitioners together around many of the themes most vital to my research. It has been a privilege to play a role in this community's growth. My future work is already indebted to the creation of this organization and the intellectual camaraderie and generosity of its members.

As I reflect on the long process of bringing this book into existence, it is a pleasure to offer special thanks to several important mentors and interlocutors who have been with this project every step of the way. Many years ago, I showed a half-baked article draft on the history of the *Overton Park* Supreme Court decision to Karen Tani. Karen, I am so glad you saw something in that messy manuscript and helped me see what it could become. The rest is history: The article won the Kathryn T. Preyer Prize (thank you, American Society for Legal History!) and launched me into this book project. Our ongoing conversations have been a source of inspiration at every turn, for this project and new ones on the horizon. And my work on *Overton Park* eventually put me into conversation with Deborah Archer—Deborah, our discussions about law and infrastructural justice have been foundational. Thank you for charting this field of research, for me and so many others, and for lighting the path toward meaningful reforms.

Meanwhile, Casey Eilbert read and commented on every word in this book, usually more than once. Thank you, Casey, for always being game to talk through archival discoveries and tease out clearer prose. I look forward to returning the favor in the near future as you finish your book. Bryan LaPointe and JP Paniagua—thanks for, well, everything. Having friends with whom to discuss the finer points of historiography and derailleur alignment in equal measures helped me keep this book on track and stay sane while doing it. José Argueta Funes, sharing drafts of our projects back and forth made a long writing process far more fun and sociable than it might have been. Anne Kerth, your insights about the political paradoxes of mobility—and so much more—made all the difference. Thank you, Casey, Bryan, JP, José, and Anne, for your dear friendship.

My final, and deepest, gratitude is to my family. My mom, dad, and brother gave me a remarkable gift: an intellectual life. To grow up in a household filled with dedication to curiosity and learning is a rare, beautiful, exhilarating thing. To share in our respective intellectual journeys as the years have gone on is a gift that keeps on giving—a gift that has made all good things in my life possible. Another remarkable gift that keeps on giving is my life with Andrea Abarca, my incredible partner, collaborator, law tutor, and hero—Andrea, thank you for every moment, thank you for every adventure, thank you for making a joyous home with me.

Notes

Introduction

1. Nell McNeil to Mike Mansfield, September 18, 1952, box 1, entry 67, Congressional Correspondence, 1951–52, Records of the Bureau of Public Roads, Record Group 30, National Archives and Records Administration, College Park, MD (hereafter cited as NARA).

2. The Bureau of Public Roads was created in 1915 within the Department of Agriculture. Its name changed to the Public Roads Administration in 1939 and moved within the Federal Works Agency. Its name reverted to the Bureau of Public Roads in 1949, when it relocated again, this time to the Department of Commerce. The Federal Highway Administration, established in 1966, eventually subsumed the BPR's functions.

3. Thomas H. MacDonald to Mike Mansfield, November 3, 1952, box 1, entry 67, Congressional Correspondence, 1951–52, Records of the Bureau of Public Roads, Record Group 30, NARA.

4. The term "citizen" appears frequently throughout this book in discussions of social equity and participatory democracy. I am not referring to the legal and exclusionary category of citizenship. Instead, I use the term to invoke the moral and political conception of citizenship, inclusive of all those existing within the territory of government under discussion, and from whom government should emanate. On this analytical and civic formulation, see Sabeel Rahman, *Democracy Against Domination* (Oxford University Press, 2016), 3–4. Other important conceptual touchstones for my consideration of citizenship include Margot Canaday, *The Straight State: Sexuality and Citizenship in Twentieth-Century America* (Princeton University Press, 2009), esp. 7–10; and, broadly, Mae M. Ngai, *Impossible Subjects: Illegal Aliens and the Making of Modern America* (Princeton University Press, 2004).

5. Federal-Aid Highway Act of 1944, Pub. Law No. 78-521, 58 Stat. 838.

6. Pub. Law No. 78-521, 58 Stat. 838; Pub. Law 84-627, 70 Stat. 374. The Highway Trust Fund followed existing trust fund models for taxation and expenditure developed during the New Deal, including the fiscal mechanism that enabled Social Security. See Julian Zelizer, *Taxing America: Wilbur D. Mills, Congress, and the State* (Cambridge University Press, 1998), 55–61. For a study of the broad relationship between petroleum-derived capital and highway construction, see Christopher Wells, "Fueling the Boom: Taxes, Invisibility, and the Growth of the American Highway Infrastructure, 1919–1956," in "Oil in American History," special issue, *Journal of American History* 99, no. 1 (2012): 72–81.

7. Jason Scott Smith, *Building New Deal Liberalism: The Political Economy of Public Works, 1933–1956* (Cambridge University Press, 2006), 31, 102.

8. I calculate present-day cost as a percentage of gross domestic product, a more representative method than consumption-based inflation metrics can provide for public works expenditure. GDP was $449.353 billion in 1956, and $27.3 trillion in 2023. See Federal Reserve Economic Data, table 1.5.5, Gross Domestic Product, Expanded Detail: Annual, https://fred.stlouisfed.org/release/tables?rid=53&eid=41630.

9. Brian Balogh, *A Government Out of Sight: The Mystery of National Authority in Nineteenth-Century America* (Cambridge University Press, 2009); and Richard White, *Railroaded: The Transcontinentals and the Making of Modern America* (W. W. Norton, 2011).

10. Gail Radford, "From Municipal Socialism to Public Authorities: Institutional Factors in the Shaping of American Public Enterprise," *Journal of American History* 90, no. 3 (2003): 863–90; Sarah L. Quinn, *American Bonds: How Credit Markets Shaped a Nation* (Princeton University Press, 2019); Destin Jenkins, *Bonded Metropolis: Debt, Redevelopment, and Racial Inequality* (University of Chicago Press, 2021).

11. "Materials in Use in U.S. Interstate Highways," United States Geological Survey, https://pubs.usgs.gov/fs/2006/3127/2006-3127.pdf.

12. Driving coast-to-coast without hitting a stoplight was a favorite talking point among the interstate highway project's boosters. See, for example, John Kelso, "Non-Stop Highways Envisioned," *Ithaca Journal*, December 17, 1956.

13. The literature touching on the interstate system's spatial impact is extensive. Among the more significant studies are Lizabeth Cohen, *A Consumer's Republic: The Politics of Mass Consumption in Postwar America* (Vintage, 2003); Owen Gutfreund, *Twentieth-Century Sprawl: Highways and the Reshaping of the American Landscape* (Oxford University Press, 2004); Kevin M. Kruse, *White Flight: Atlanta and the Making of Modern Conservatism* (Princeton University Press, 2005); Paul Sutter, *Driven Wild: How the Fight Against Automobiles Launched the Modern Wilderness Movement* (University of Washington Press, 2002); and Chris Wells, *Car Country: An Environmental History* (University of Washington Press, 2012).

14. Robert Moses, "The New Super-Highways: Blessing or Blight?" *Harper's Magazine*, December 1956, 31.

15. On "the administrative ordering of nature and society" and the general concept of high modernism see James C. Scott, *Seeing Like a State: How Certain Schemes to Improve the Human Condition Have Failed* (Yale University Press, 1998), 4; for further exploration of high modernism and the significance of 1945, see James C. Scott, *The Art of Not Being Governed: An Anarchist History of Upland Southeast Asia* (Yale University Press, 2009), xii.

16. Timothy Mitchell, *Carbon Democracy: Political Power in the Age of Oil* (Verso, 2011), 134–9, 176.

17. On the rise of the federal administrative state and its characterizations in the mid-twentieth century, see Anne Kornhauser, *Debating the American State: Liberal Anxieties and the New Leviathan, 1930–1970* (University of Pennsylvania Press, 2015), 1; and Karen Tani, *States of Dependency: Welfare, Rights, and American Governance* (Cambridge University Press, 2016), 21. On the long rise of administrative governance in the United States, see William Novak, *New Democracy: The Creation of the Modern American State* (Harvard University Press, 2022). Of course, states remained powerful subnational governing units despite—and sometimes because of—federal action throughout the twentieth century. See Tani, *States of Dependency*; and Brent Cebul, Karen Tani, and Mason B. Williams, "Clio and the Compound Republic," *Publius: The Journal of Federalism* 47, no. 2 (2017): 235–59. Meanwhile, the aims and powers of administrative governance have been studied with respect to a wide range of specific policy areas, ranging

across issues including everything from managing banks to provisioning welfare to policing gender and sexuality. See, for example, Peter Conti-Brown and Sean Vanatta, *Private Finance, Public Power: A History of Bank Supervision in America* (Princeton University Press, 2025); Linda Gordon, *Pitied but Not Entitled: Single Mothers and the History of Welfare* (Harvard University Press, 1994); and Canaday, *Straight State.*

18. On the origins of the American administrative state, see Novak, *New Democracy.* On the administrative state's ability—at least in theory—to check private power and protect the public interest on behalf of all citizens, see Rahman, *Democracy Against Domination.*

19. On "sympathetic critics" and the range of problems they decried, see Kornhauser, *Debating the American State.*

20. Simon Kuznets articulated his concerns in "National Income, 1929–1932," Senate Document 124, January 4, 1934, 7.

21. John D. Morris, "Eisenhower Signs Road Bill; Weeks Allocates $1.1 Billion," *New York Times*, June 30, 1956; Joseph C. Ingraham, "U.S. Drivers Begin Footing New Highway Bill," *New York Times*, July 1, 1956; and Joseph C. Ingraham, "U.S. Gasoline Tax Up a Penny Today," *New York Times*, July 1, 1956.

22. "Scars Across the Land," *New York Times*, July 10, 1968.

23. Ben Kelley and Richard Hébert, "Priorities or Trust Funds? *The Nation*, April 19, 1971, 497.

24. Deborah Archer, *Dividing Lines: How Transportation Infrastructure Reinforces Racial Inequality* (W. W. Norton, 2025), 24.

25. Archer, *Dividing Lines*, 6, 10, 12, 24; Francesca Russello Ammon, *Bulldozer: Demolition and the Clearance of the Postwar Landscape* (Yale University Press, 2016), 183; Nathan D. B. Connolly, *A World More Concrete: Real Estate and the Remaking of Jim Crow South Florida* (University of Chicago Press, 2014), 8, 183. See also Ryan Reft, Amanda Phillips de Lucas, and Rebecca Retzlaff, eds., *Justice and the Interstates: The Racist Truth About Urban Highways* (Island Press, 2023); and Joseph DiMento and Cliff Ellis, *Changing Lanes: Visions and Histories of Urban Freeways* (MIT Press, 2013).

26. On the emergence of "freeway fighters," see Raymond Mohl, "Stop the Road: Freeway Revolts in American Cities," *Journal of Urban History* 30, no. 5 (2004): 674–706. For details on San Francisco, see Richard F. Weingroff, "The Greatest Decade, 1956–1966," part 1, "Essential to the National Interest," Highway History, Federal Highway Administration, updated June 30, 2023, https://www.fhwa.dot.gov/infrastructure/50interstate.cfm. On Washington Square Park, see Eric Avila, *The Folklore of the Freeway: Race and Revolt in the Modernist City* (University of Minnesota Press, 2014), 53–73.

27. *Nashville I-40 Steering Committee v. Ellington*, 390 U.S. 921 (1968); *Citizens to Preserve Overton Park, Inc. v. Volpe*, 401 U.S. 402 (1971).

28. Regarding the development of twentieth-century American administrative law and governance and its vexed relationship with liberalism and participatory democracy, see Daniel R. Ernst, *Tocqueville's Nightmare: The Administrative State Emerges in America, 1900–1940* (Oxford University Press, 2014); Joanna Grisinger, *The Unwieldy American State: Administrative Politics Since the New Deal* (Cambridge University Press, 2012); and Kornhauser, *Debating the American State.* On the legal history of administrative governance and its relationship with the citizenry and with the judiciary, see Reuel Schiller, "Enlarging the Administrative Polity: Administrative Law and the Changing Definition of Pluralism, 1945–1970," *Vanderbilt Law Review* 53, no. 5 (2000): 1389–1454; and Reuel Schiller, "'Saint George and the Dragon': Courts and the

Development of the Administrative State in Twentieth-Century America," *Journal of Policy History* 17, no. 1 (2005): 110–24.

29. Vera Candiani, *Dreaming of Dry Land: Environmental Transformation in Colonial Mexico City* (Stanford University Press, 2014), xvi.

30. Michael Mann, "The Autonomous Power of the State: Its Origins, Mechanisms, and Results," in *States in History*, ed. John A. Hall (Oxford University Press, 1986), 109, 113. Mann contrasted "despotic power" and "infrastructural power." Despotic power referred to the range of official activities a state could undertake without engaging in negotiation with the populace. Infrastructural power dealt with a state's ability to undertake action, whether or not the action in question was despotic. Mann published an initial version of this work in 1984; see Michael Mann, "The Autonomous Power of the State: Its Origins, Mechanisms and Results." *European Journal of Sociology / Archives Européennes de Sociologie / Europäisches Archiv Für Soziologie* 25, no. 2 (1984): 185–213. For discussions of some of the limitations of Mann's theorizations, see Philip Gorski, "Mann's Theory of Ideological Power: Applications and Elaborations," in *An Anatomy of Power: The Social Theory of Michael Mann*, ed. John A. Hall and Ralph Schroeder (Cambridge University Press, 2006); and Gabriel N. Rosenberg, "Youth as Infrastructure: 4-H and the Intimate State in 1920s Rural America," in *The Boundaries of the State in U.S. History*, ed. James T. Sparrow, William Novak, and Stephen W. Sawyer (University of Chicago Press, 2015). William Novak has contributed perhaps the most comprehensive treatment to date of Mann's theory of infrastructural power; see "The Myth of the 'Weak' American State," *American Historical Review* 113, no. 3 (2008): 752–72.

31. Novak, "Myth of the 'Weak' American State," 763.

32. Mary Bridges, "The Infrastructural Turn in Historical Scholarship," *Modern American History* 6, no. 1 (2023): 103–20; David Alff and Jo Guldi, "Infrastructure: A Useful Anachronism?" *Eighteenth-Century Studies* 58, no. 1 (2024): 1–9. Of course, analysis of infrastructure has long been of interest to historians, even if the recent "infrastructural turn" has made it a more self-conscious area of study. Spanning generations of scholarship and varied areas of emphasis, significant touchstones include Candiani, *Dreaming of Dry Land*; Connolly, *A World More Concrete*; Robin Einhorn, *Property Rules: Political Economy in Chicago, 1833–1872* (University of Chicago Press, 1991); Julie Greene, *The Canal Builders: Making America's Empire at the Panama Canal* (Penguin Press, 2009); Thomas Hughes, *Networks of Power: Electrification in Western Society* (Johns Hopkins University Press, 1983); Katherine Johnson, *The American Road: Highways and American Political Development, 1891–1956* (University of Kansas Press, 2021); Martin Melosi, *The Sanitary City: Urban Infrastructure in American from Colonial Times to the Present* (Johns Hopkins University Press, 2000); Andrew Needham, *Power Lines: Phoenix and the Making of the Modern Southwest* (Princeton University Press, 2014); Gail Radford, *The Rise of the Public Authority: Statebuilding and Economic Development in Twentieth-Century America* (University of Chicago Press, 2013); Gabriel Rosenberg, *4-H Harvest: Sexuality and the State in Rural America* (University of Pennsylvania Press, 2016); Paul Sabin, *Crude Politics: The California Oil Market, 1900–1940* (University of California Press, 2005); Kyle Shelton, *Power Moves: Transportation, Politics, and Development in Houston* (University of Texas Press, 2017); Carol Sheriff, *The Artificial River: The Erie Canal and the Paradox of Progress, 1817–1862* (Hill and Wang, 1996); George Rogers Taylor, *The Transportation Revolution, 1815–1860* (Rinehart, 1951); Joel Tarr, *The Search for the Ultimate Sink: Urban Pollution in Historical Perspective* (University of Akron Press, 1996); and White, *Railroaded*.

33. White House Briefing Room, "Fact Sheet: The American Jobs Plan," press release, March 31, 2021, https://bidenwhitehouse.archives.gov/briefing-room/statements-releases/2021/03/31/fact-sheet-the-american-jobs-plan/.

34. April Ryan, "Buttigieg Says Racism Built into US Infrastructure Was a 'Conscious Choice,'" *The Grio*, April 6, 2021, https://thegrio.com/2021/04/06/pete-buttigieg-racism-us-infrastructure/; Sydney Pereira, "Buttigieg Observed Some Highways Have a Racist Past. Here's What That Meant In NYC," *The Gothamist*, April 11, 2021, https://gothamist.com/news/buttigieg-observed-some-highways-have-racist-past-heres-what-meant-nyc; Corinne Grinapol, "Biden Administration Seeks to Address the Interstate Highway System's Racist Past," *Engineering News-Record*, April 15, 2021, https://www.enr.com/articles/51593-biden-administration-seeks-to-address-the-interstate-highway-systems-racist-past.

35. Reinhold Martin, *The Urban Apparatus: Mediapolitics and the City* (University of Minnesota Press, 2016), 1, 142.

36. Deborah Archer and Joseph Schottenfeld, "Defending Home: Toward a Theory of Community Equity," November 14, 2025, available at SSRN, https://papers.ssrn.com/sol3/papers.cfm?abstract_id=4971493.

37. This resistance has only continued to build, up to the present day. As legal scholar Sabeel Rahman writes, "we live in an oddly undemocratic time," in which power tends to flow to "the market" and to "experts, judges, or political elites." Sometimes these two categories of power are arrayed against one another; sometimes they are a unity. Either way, direct citizen power remains subdued. See *Democracy Against Domination*, 4.

38. Brent Cebul, *Illusions of Progress: Business, Poverty, and Liberalism in the American Century* (University of Pennsylvania Press, 2023), 8–11. On the broader phenomenon of mid-twentieth-century suppression of citizen power in government, see Teal Arcadi, "Undemocracy in America: Excavating the Foundations of Political Domination in the Modern United States," *Reviews in American History* 52, no. 1 (2024): 75–93.

39. Paul Sabin, *Public Citizens: The Attack on Big Government and the Remaking of American Liberalism* (W. W. Norton, 2021).

40. Rachel Carson, for example, spoke about "authoritarian control" with regard to the agencies in a 1963 address to the New England Wildflower Preservation Society, as quoted in Sabin, *Public Citizens*, 20.

41. Ralph Nader, "Ombudsmen for State Governments," in *The Ombudsman: Citizen's Defender*, ed. Donald C. Rowat (Allen & Uwin, 1965), 240. See also Sabin, *Public Citizens*, 26–27.

42. Casey Eilbert, "Conceptualizing the 'Iron Cage': Bureaucracy in Modern America" (PhD diss., Princeton University, 2024), chap. 6, "Government Bureaucracy."

43. See, generally, Archer and Schottenfeld, "Defending Home."

44. Deborah N. Archer, "'White Men's Roads Through Black Men's Homes': Advancing Racial Equity Through Highway Reconstruction," *Vanderbilt Law Review* 73, no. 5 (2020), 1314–21; see also Archer and Schottenfeld, "Defending Home."

45. On the CFPB's structure, see Sabeel Rahman and Hollie Russon Gilman, *Civic Power: Rebuilding American Democracy in an Era of Crisis* (Cambridge University Press, 2019), 143–50. For an important example of the CFPB's ability to withstand anti-democracy lawsuits, see *Consumer Financial Protection Bureau v. Community Financial Services Association of America, Ltd.*, 601 U.S. 416 (2024). The second Trump administration attempted to use executive authority to decimate the CFPB's staffing in early 2025; at least initially, federal courts blocked that effort. See Dylan Tokar, "Federal Judge Orders a Halt to Trump Administration Cuts at CFPB," *Wall Street Journal*, February 14, 2025.

46. Rahman and Gilman, *Civic Power*, 148; Laurel Wamsley and Rafael Nam, "Federal Judge Blocks CFPB from Laying off More Employees," NPR, February 14, 2025, https://www

.npr.org/2025/02/14/nx-s1-5297870/cfpb-doge-trump-musk-federal-workforce; Greg Iacurci, "Trump's 'Big Beautiful Bill' Slashes CFPB Funding: What It Means for You," CNBC, July 9, 2025, https://www.cnbc.com/2025/07/09/trump-big-beautiful-bill-slashes-cfpb-funding-what-it-means.html.

47. *CFPB v. Community Financial Services Association of America, Ltd.*; *Loper Bright Enterprises v. Raimondo*, 603 U.S. 369 (2024). Other Supreme Court rulings announced in 2024 weakened other dimensions of administrative governance. In *Corner Post, Inc. v. Board of Governors of the Federal Reserve System*, 603 U.S. 799 (2024), the Supreme Court's conservative wing broadened the statute of limitations regarding when judicial review of administrative action can take place, making it significantly easier for judges to make determinations about administrative action. In *Securities and Exchange Commission v. Jarkesy*, 603 U.S. 109 (2024), they broke with 170 years of precedent to rule that a defendant accused of breaching a federal agency's regulations is entitled to a jury trial in federal court—functionally stripping agencies of enforcement power.

48. The "blank check" commentary came from the legal scholar Sabeel Rahman at a Roosevelt Institute panel discussion titled "Preserving the Administrative State: Threats to Administrative Law Enforcement in the Courts," April 9, 2024. His comment is time-stamped at https://youtu.be/Q3QUKuaWqhI?si=VCYXCS_ILfFKG7PB&t=1977. Other legal scholars have questioned how seismic the *Loper Bright* ruling really will be. Partly because the federal courts do not have the capacity to wade into every administrative issue, change may not be as severe as some fear. See Cary Coglianese and Daniel E. Walters, "The Great Unsettling: Administrative Governance After *Loper Bright*," December 7, 2024, available at SSRN, https://papers.ssrn.com/sol3/papers.cfm?abstract_id=5046620; and Cass Sunstein, "The Consequences of *Loper Bright*," July 8, 2024, available at SSRN, https://papers.ssrn.com/sol3/papers.cfm?abstract_id=4881501.

49. Nell McNeil to Mike Mansfield, September 18, 1952, box 1, entry 67, Congressional Correspondence, 1951–52, Records of the Bureau of Public Roads, Record Group 30, NARA.

Chapter One

1. "California Highways and Public Works," *California Highways and Public Works* (1936): 1, 33; "It Actually Rained," *San Francisco Examiner*, December 8, 1936. La Follette remained on the schedule until at least December 6, 1936; see "Highways," *San Francisco Examiner*, December 6, 1936.

2. Presidential election statistics for 1936, American Presidency Project, https://www.presidency.ucsb.edu/statistics/elections/1936.

3. John Maynard Keynes, *The General Theory of Employment, Interest and Money* (1936; Wordsworth, 2017). On the Keynesian revolution and macroeconomic shifts in the second half of the 1930s, see Alan Brinkley, *The End of Reform: New Deal Liberalism in Recession and War* (Alfred A. Knopf, 1995), 6–8, 31–47, 128–31. On conservative attacks on the New Deal, see Kim Phillips-Fein, *Invisible Hands: The Making of the Conservative Movement from the New Deal to Reagan* (W. W. Norton, 2009).

4. H. S. Fairbank, "Objects and Methods of the State-Wide Highway Planning Surveys," American Association of State Highway Officials Annual Convention, San Francisco, California, December 8, 1936, box 1, entry 12-A, Office Files and Correspondence Concerning Particular Bureau Activities, Records of the Bureau of Public Roads, Record Group 30, National Archives and Records Administration, College Park, MD (hereafter cited as NARA), 5, 7.

5. Christopher W. Wells, *Car Country: An Environmental History* (University of Washington Press, 2021), 81–82; David J. St. Clair, *The Motorization of American Cities* (Praeger, 1986), 19–20;

Richard John, *Spreading the News: The American Postal System from Franklin to Morse* (Harvard University Press, 1995).

6. On the long American tradition of keeping the federal government "out of sight," see Brian Balogh, *A Government Out of Sight: The Mystery of National Authority in Nineteenth-Century America* (Cambridge University Press, 2009). On the twentieth-century iteration of this tradition, see Brent Cebul and Mason B. Williams, "'Really and Truly a Partnership': The New Deal's Associational State and the Making of Postwar American Politics," in *Shaped by the State: Toward a New Political History of the Twentieth Century*, ed. Brent Cebul, Lily Geismer, and Mason B. Williams (University of Chicago Press, 2019), 96–122; and Brian Balogh, *The Associational State: American Governance in the Twentieth Century* (University of Pennsylvania Press, 2015). A key study of how state action has been submerged and hidden in associational relationships and market mechanisms is Gail Radford, *The Rise of Public Authority: Statebuilding and Economic Development in Twentieth-Century America* (University of Chicago Press, 2013).

7. Pub. Law No. 64-156, 39 Stat. 355.

8. The highway official is quoted in St. Clair, *Motorization of American Cities*, 22.

9. St. Clair, *Motorization of American Cities*, 23.

10. Fairbank, "Objects and Methods," 3, 6.

11. Fairbank, 2.

12. Fairbank, 3–4, 7. In 1920, the US Census reported that 48.8 percent of the population lived in rural areas. By 1930, that number had decreased to 43.9 percent. In the 1910 census—reflecting a national context in which highway policy had emphasized farm-to-market transportation—the rural population had accounted for 54.4 percent of the national population. See "Urban and Rural Population: 1900–1990," US Census Bureau, October 1995, https://www2.census.gov/programs-surveys/decennial/tables/1990/1990-urban-pop/urpop0090.txt.

13. Keynes, *General Theory*, xvi.

14. Otis Graham, *Losing Time: The Industrial Policy Debate* (Harvard University Press, 1992), 65; Timothy Mitchell, "Fixing the Economy," *Cultural Studies* 12, no. 1 (1998): 90–91.

15. See Brinkley, *End of Reform*, 6–8, 31–47, 128–31.

16. Keynes, *General Theory*, xvi; Timothy Mitchell, *Carbon Democracy: Political Power in the Age of Oil* (Verso, 2011), 136–37.

17. Mitchell, *Carbon Democracy*, 137; Jonathan Levy, *The Real Economy: History and Theory* (Princeton University Press, 2025), 195.

18. Mitchell, "Fixing the Economy," 90–91.

19. Fairbank, "Objects and Methods," 3; President's Advisory Committee on a National Highway Program, *A Ten-Year National Highway Program* (US Government Printing Office, 1955), 29.

20. Wells, *Car Country*, 37–38; "Making Automobiles Last During WWII," National World War II Museum, January 6, 2022, https://www.nationalww2museum.org/war/articles/automobile-rationing-world-war-ii. The sale of new automobiles rose from 181,000 in 1910 to 4.5 million in 1929—about one new car for every twenty-seven people. See https://www.richmondfed.org/publications/research/econ_focus/2003/winter/economic_history.

21. The BPR and *Fortune* are quoted in Wells, *Car Country*, 132.

22. "Motor Train Headed Here," *Los Angeles Times*, June 29, 1919; Owen D. Gutfreund, *Twentieth-Century Sprawl: Highways and the Reshaping of the American Landscape* (Oxford University Press, 2004), 23–24; Ann Manchester and Albert Manchester, "From DC to the Golden Gate: The First Transcontinental Military Motor Convoy," *American History* 32, no. 5 (1997): 39.

23. Eisenhower is quoted in Manchester and Manchester, "From DC to the Golden Gate," 39.

24. E. Y. Watson, "Truck Convoy Doubled for Journey Across America," *Detroit Free Press*, June 22, 1919; "Ohio Welcomes Touring Trucks," *Washington Post*, July 11, 1919; "Army Convoy Rolls Along on Schedule," *Detroit Free Press*, July 13, 1919.

25. Sheppard Butler, "Exhaust Echoes," *Chicago Tribune*, July 19, 1919.

26. Manchester and Manchester, "From DC to the Golden Gate," 39, 42, 69.

27. US War Department, *Annual Reports, 1920*, vol. 1 (Government Printing Office, 1921), 1789.

28. Wells, *Car Country*, 185.

29. Wells, 185–86; Oregon Department of Transportation, "Fuels Tax History," accessed December 9, 2025, https://www.oregon.gov/odot/ftg/pages/about-us.aspx; American Society of Civil Engineers, "Columbia River Scenic Highway," accessed December 9, 2025, https://www.asce.org/about-civil-engineering/history-and-heritage/historic-landmarks/columbia-river-scenic-highway.

30. Wells, *Car Country*, 174–75, 185–88.

31. Wells, 187–88. The quote from the Tennessee tax collector is at 187.

32. On the infrastructural implications of the New Deal's approach to state-sponsored macroeconomic management via relief and public works, see Jason Scott Smith, *Building New Deal Liberalism: The Political Economy of Public Works, 1933–1956* (Cambridge University Press, 2006).

33. Herbert Hoover, "Message to the American Road Builders Association," January 11, 1932, American Presidency Project, https://www.presidency.ucsb.edu/documents/message-the-american-road-builders-association.

34. Wells, *Car Country*, 188.

35. Wells, 189.

36. US Department of Agriculture, Press Service, "$400,000,000 Road Fund Apportioned Among States," June 23, 1933, box 8, entry 12-G, Office Files and Correspondence Concerning Particular Bureau Activities, Records of the Bureau of Public Roads, Record Group 30, NARA.

37. Harold L. Ickes, *The Secret Diaries of Harold L. Ickes*, vol. 1, *The First Thousand Days 1933–1936* (Da Capo Press, 1974), 243–44.

38. Ickes, *Secret Diaries*, 243–44.

39. Newspapers often remarked on Roosevelt's personal interest in the process. When Roosevelt drew up his own plans, he "displayed what friends described as 'uncanny knowledge' of American geography," reported Douglas Warrenfels in "Super-Roads Considered in Work Plan," *Washington Post*, February 20, 1935.

40. Diaries of Henry Morgenthau Jr., December 10, 1934, Franklin D. Roosevelt Presidential Library and Museum, http://www.fdrlibrary.marist.edu/archives/collections/franklin/index.php?p=collections/findingaid&id=535&q=&rootcontentid=188897#id188897.

41. This fit the trend of channeling federal funding through local agencies. Generally, see Cebul and Williams, "'Really and Truly a Partnership'"; and Brent Cebul, *Illusions of Progress: Business, Poverty, and Liberalism in the American Century* (University of Pennsylvania Press, 2023).

42. Letter to Roosevelt, May 23, 1933, quoted in Tom Lewis, *Divided Highways: Building The Interstate Highways, Transforming American Life* (1997; Cornell University Press, 2013), 22.

43. Mark H. Rose and Raymond Mohl, *Interstate: Highway Politics and Policy Since 1939* (1979; University of Tennessee Press, 2012), 5.

44. Works Progress Administration, "Summary of Physical Accomplishments on Selected Types of WPA Projects," December 11, 1936, box 8, entry 12-G, Office Files and Correspondence

Concerning Particular Bureau Activities, Records of the Bureau of Public Roads, Record Group 30, NARA.

45. Wells, *Car Country*, 192–93.

46. Wells, 175, 192–93.

47. Warrenfels, "Super-Roads Considered in Work Plan."

48. Fairbank, "Objects and Methods," 3, 5, 11.

49. Wells, *Car Country*, 194.

50. Richard White, *Railroaded: The Transcontinentals and the Making of Modern America* (W. W. Norton, 2012), 2–9.

51. Wells, *Car Country*, 194.

52. National Resource Committee, "Our Cities: Their Place in the National Economy" (US Government Printing Office, 1937), v–ix, 1–3, 41; Emily A. Prifogle, "Cows, Cars, and Criminals: The Legal Landscape of the Rural Midwest, 1920–1975" (PhD diss., Princeton University, 2019), 11.

53. Federal-Aid Highway Act of 1944, Pub. Law No. 78-521, 58 Stat. 838. More than a decade later, Congress delivered funding sufficient to construct the interstate highway system with the Federal-Aid Highway Act and Highway Revenue Act of 1956, Pub. Law 84-627, 70 Stat. 374.

54. Thomas MacDonald, "Looking Toward the Highway Future," American Association of State Highway Officials Annual Convention, San Francisco, California, December 8, 1936, box 1, entry 12-A, Office Files and Correspondence Concerning Particular Bureau Activities, Records of the Bureau of Public Roads, Record Group 30, NARA, 2–3.

55. Retrospectively proving MacDonald's point, the larger M4 now parallels the A4 in England.

56. MacDonald, "Looking Toward the Highway Future," 5.

57. MacDonald, 7–11.

58. MacDonald, 7–11. On the German economy in this period, see Adam Tooze, *The Wages of Destruction: The Making and Breaking of the Nazi Economy* (Allen Lane, 2006), esp. chaps. 6 and 7.

59. MacDonald, "Looking Toward the Highway Future," 11–12.

60. MacDonald, 12–13.

61. US Department of Agriculture, Press Service, "Highways Will Be Safer: Driver Remains A Problem," December 31, 1936, box 8, entry 12-G, Office Files and Correspondence Concerning Particular Bureau Activities, Records of the Bureau of Public Roads, Record Group 30, NARA.

62. MacDonald, "Looking Toward the Highway Future," 13–14.

63. US Department of Agriculture, Press Service, "Research Reveals Better Ways of Building Low-Cost Roads," June 23, 1936, box 8, entry 12-G, Office Files and Correspondence Concerning Particular Bureau Activities, Records of the Bureau of Public Roads, Record Group 30, NARA; US Department of Agriculture, Press Service, "Public Roads Bureau Gives Course in Study of Soils," June 15, 1937, box 8, entry 12-G, Office Files and Correspondence Concerning Particular Bureau Activities, Records of the Bureau of Public Roads, Record Group 30, NARA.

64. US Department of Agriculture, Press Service, "Pickup of Sound Waves Locates Underground Rock," March 18, 1936, box 8, entry 12-G, Office Files and Correspondence Concerning Particular Bureau Activities, Records of the Bureau of Public Roads, Record Group 30, NARA.

65. US Department of Agriculture, Press Service, "Candid Camera Shows Need for Ample Highway Width," October 6, 1937, box 8, entry 12-G, Office Files and Correspondence Concerning Particular Bureau Activities, Records of the Bureau of Public Roads, Record Group 30, NARA.

66. US Department of Agriculture, Press Service, "36 States Survey Highways as Basis for Road Programs," June 18, 1936, box 8, entry 12-G, Office Files and Correspondence Concerning Particular Bureau Activities, Records of the Bureau of Public Roads, Record Group 30, NARA.

67. US Department of Agriculture, Press Service, "Wallace Names 12 Highway Experts to Work with Public Roads Bureau," May 22, 1937, box 8, entry 12-G, Office Files and Correspondence Concerning Particular Bureau Activities, Records of the Bureau of Public Roads, Record Group 30, NARA.

68. US Department of Agriculture, Press Service, "Contracts Awarded for Research Station for Bureau of Public Roads," October 27, 1937, box 8, entry 12-G, Office Files and Correspondence Concerning Particular Bureau Activities, Records of the Bureau of Public Roads, Record Group 30, NARA.

69. Rose and Mohl, *Interstate*, 5.

70. MacDonald is quoted in Rose and Mohl, *Interstate*, 5.

71. US Department of Agriculture, Press Service, "Surfaced Roads Now Reach Almost All Parts of US," December 4, 1938, box 8, entry 12-G, Office Files and Correspondence Concerning Particular Bureau Activities, Records of the Bureau of Public Roads, Record Group 30, NARA.

72. Bureau of Public Roads, *Toll Roads and Free Roads* (US Government Printing Office, 1939), vii, ix.

73. Bureau of Public Roads, *Toll Roads and Free Roads*, 1–5, 13–15, 89; Gallup Poll #1938-0113: War Debts / Automobiles / Roosevelt, Gallup Organization (Roper Center for Public Opinion Research, Cornell University, 1938), dataset, https://doi.org/10.25940/ROPER-31087097.

74. Wells, *Car Country*, 198. For a more detailed study of the New York State Thruway, see Michael Fein, *Paving the Way: New York Road Building and the American State, 1880–1956* (University Press of Kansas, 2008).

75. St. Clair, *Motorization of American Cities*, 132.

76. Bureau of Public Roads, *Toll Roads and Free Roads*, 95.

77. Bureau of Public Roads, vii, 89–90, 122; see also Wells, *Car Country*, 197.

78. Bureau of Public Roads, *Toll Roads and Free Roads*, 95–97.

79. Wells, *Car Country*, 196.

80. Bureau of Public Roads, *A Report of the Status and Activities of the Highway Traffic Advisory Committee to the War Department as of December 31, 1942*, box 831, entry 54, Highway Traffic Advisory Committee to the War Department, 1941–1945—Minutes of the Committee and Press Releases, Records of the Bureau of Public Roads, Record Group 30, NARA.

81. Roosevelt is quoted in Rose and Mohl, *Interstate*, 16.

82. G. Donald Kennedy, "Relation of Federal Defense Highway Appropriations and the States' Regular Aid Programs," *Proceedings of the American Association of State Highway Officials* (American Association of State Highway Officials, 1941), 1.

83. Kennedy, "Relation of Federal Defense Highway Appropriations," 4–5.

84. Kennedy, 1, 3–4.

85. Kennedy, 4.

86. Kennedy, 4; The Ford Motor Company's involvement is striking and demands further research. This episode's bearing on the associational qualities of highway-borne state building is evident, particularly with regard to the automotive industry's involvement in eventual

interstate planning. On Ford's Ypsilanti Plant, see the Ypsilanti Historical Society, "Ford Ypsilanti Plant—1932," http://www.ypsilantihistoricalsociety.org/history/page999995.html.

87. Kennedy, "Relation of Federal Defense Highway Appropriations," 4. Today, Interstate 94 joins Detroit to Chicago, and arterial routes connect metropolises like Ypsilanti, Ann Arbor, and Kalamazoo to the main highway across the state.

88. Franklin D. Roosevelt, Executive Order 8989 Establishing the Office of Defense Transportation, https://www.presidency.ucsb.edu/documents/executive-order-8989-establishing-the-office-defense-transportation; see also John L. Rogers, Central Motor Transportation Committee, PRA, message on Office of Defense Transportation, box 3, entry 6-C, General Correspondence and Related Records, 1951–55, Records of the Bureau of Public Roads, Record Group 30, NARA; Roosevelt signed the Defense Highway Act into law on November 19, 1941, Pub. Law 77-295, 55 Stat. 765.

89. Pub. Law 77-295, 55 Stat. 765.

90. Thomas H. MacDonald, "The City's Place in Post-War Highway Planning: Concrete Advice on What Cities Should Do to Help Provide the Economic 'Backlog' Against Post-War Recession," *American City*, February 1943, 42.

91. MacDonald, "City's Place in Post-War Highway Planning," 42.

92. Franklin D. Roosevelt to Thomas H. MacDonald, April 14, 1941, box 3, entry 6-C, General Correspondence and Related Records, 1951–55, Records of the Bureau of Public Roads, Record Group 30, NARA.

93. Franklin D. Roosevelt to John M. Carmody, April 14, 1941, box 3, entry 6-C, General Correspondence and Related Records, 1951–55, Records of the Bureau of Public Roads, Record Group 30, NARA.

94. Fears of a postwar return to economic recession or depression shaped immediate postwar policymaking and crystallized in the Employment Act of 1946. See Stephen Kemp Bailey, *Congress Makes a Law: The Story Behind the Employment Act of 1946* (Vintage Books, 1964).

95. H. S. Fairbank, "Post-War Planning for Highway Construction," box 822, entry 54, Highway Traffic Advisory Committee to the War Department, 1941–45: Director's Correspondence—State and Local Administrators, Records of the Bureau of Public Roads, Records Group 30, NARA. The transcript of Fairbank's remarks is undated, and his audience is unidentified. His reference in the text to 1936 state highway planning surveys "six years past," however, indicates he delivered the address sometime in 1942, and his references to raw materials, concrete production, and construction techniques reveal this is one of the many addresses PRA officials delivered to representatives of the road-building industry throughout the 1940s.

96. Fairbank, "Post-War Planning for Highway Construction," n.p.

97. Fairbank, n.p.

98. Bureau of Public Roads, *Interregional Highways* (US Government Printing Office, 1944), 1.

99. Bailey, *Congress Makes a Law*, 3–36.

100. "Statement by the President," December 20, 1944, box 9, entry 12-G, Office Files and Correspondence Concerning Particular Bureau Activities, Records of the Bureau of Public Roads, Record Group 30, NARA.

101. Pub. Law 78-521, 58 Stat. 838; St. Clair, *Motorization of American Cities*, 150.

102. Previously, funding for urban areas had been entirely at the discretion of individual states. The majority of the states directed little, if any, federal funding into urban areas. See St. Clair, *Motorization of American Cities*, 149.

103. Pub. Law 78-521, 58 Stat. 838; Wells, *Car Country*, 270.

104. Pub. Law 78-521, 58 Stat. 838.

105. Federal Works Agency, "For Release," August 3, 1947, box 9, entry 12-G, Office Files and Correspondence Concerning Particular Bureau Activities, Records of the Bureau of Public Roads, Record Group 30, NARA. See also Rose and Mohl, *Interstate*, 31; and Wells, *Car Country*, 271–72.

Chapter Two

1. Richard F. Weingroff, "President Eisenhower's Big Day," Highway History, Federal Highway Administration, updated June 27, 2017, https://www.fhwa.dot.gov/infrastructure/bigday.cfm; Tom Lewis, *Divided Highways: Building the Interstate Highways, Transforming American Life* (Cornell University Press, 2013), xiii, 121–23; Eric Avila, *The Folklore of the Freeway: Race and Revolt in the Modernist City* (University of Minnesota Press, 2014), 1.

2. Take, for example, Glenda Gilmore and Thomas Sugrue's framing of the interstate highway project in their widely read textbook: "Eisenhower . . . launched the largest public works program in American history, the construction of the interstate highway system." While they acknowledge the presence of earlier federal highway aspirations, they make no mention of the prefiguring Federal-Aid Highway Act of 1944, Pub. Law 78-521, 58 Stat. 838 (1944), nor the centrality of Congress in the interstate highway development process. Gilmore and Sugrue, *These United States: A Nation in the Making, 1890 to the Present* (W. W. Norton, 2015), 346. Eric Foner used a nearly identical framing in his textbook *Give Me Liberty: An American History* (W. W. Norton, 2015), 948. Meanwhile, the United States Army continued to claim that nuclear defense was a key factor in the interstate system's design; see Lee Lacy, "Dwight D. Eisenhower and the Birth of the Interstate System," February 20, 2018, United States Army, https://www.army.mil/article/198095/dwight_d_eisenhower_and_the_birth_of_the_interstate_highway_system. Along the way, subsequent presidents propped up the Eisenhower-centric narrative to advance their own agendas. Ronald Reagan, for example, declared a "National Interstate Highway Day" on the thirtieth anniversary of the 1956 highway act, invoking Eisenhower's leadership and calling the interstate system "the world's largest and most successful transportation and public works project" as he pushed for his own infrastructure plans. Not long after, George H. W. Bush more blatantly attached Eisenhower to the nation's defining infrastructure system when he redesignated it as the "Dwight D. Eisenhower National System of Interstate and Defense Highways." And a generation later, Barack Obama told the nation that when Eisenhower "built the interstates" he showed that "infrastructure should not be a partisan issue." Had Obama looked a little closer, he might have discovered the interstate system's deeper and more partisan legislative foundations and drawn different lessons from its history. On subsequent presidential invocations, see Richard F. Weingroff, "In Memory of Ronald Reagan," Highway History, Federal Highway Administration, updated June 27, 2017, https://www.fhwa.dot.gov/infrastructure/reagan.cfm; Ronald Reagan, "Proclamation 5503, National Interstate Highway Day (1986)," online by Gerhard Peters and John T. Woolley, American Presidency Project, https://www.presidency.ucsb.edu/node/258926; Richard F. Weingroff, "A Bush at Both Ends: Before and After the Interstate Era," Highway History, Federal Highway Administration, updated June 27, 2017, https://www.fhwa.dot.gov/infrastructure/rw01d.cfm; "An Act to Redesignate the National System of Interstate and Defense Highways as the Dwight D. Eisenhower National System of Interstate and Defense Highways," Pub. Law 101-427, 104 Stat. 927 (1990); David Hudson, "President Obama: 'Let's Build Some Bridges, Let's Build Some Roads,'" July 17, 2014, *White House Blog*, https://

obamawhitehouse.archives.gov/blog/2014/07/17/president-obama-lets-build-some-bridges-lets-build-some-roads; and Philip Bump, "Obama Calls for Infrastructure Spending—for the Fifth Time in Five Years," *The Atlantic*, March 29, 2013, https://www.theatlantic.com/politics/archive/2013/03/obama-calls-infrastructure-spending-fifth-time-five-years/316815/.

3. Federal-Aid Highway Act of 1944, Pub. Law 78-521, 58 Stat. 838; Federal-Aid Highway Act of 1956, Pub. Law 84-627, 70 Stat. 374; Weingroff, "President Eisenhower's Big Day"; Lewis, *Divided Highways*, 121.

4. Historians have recognized but not emphasized partisanship in the interstate highway system's origins. See Mark H. Rose and Raymond A. Mohl, *Interstate: Highway Politics and Policy since 1939* (1979; University of Tennessee Press, 2012), 29–40, 87–94. This analysis emphasizes a period prior to the partisan congressional debate over bonds and taxes that developed in 1955 and details how Congress interacted with the executive branch and lobbying efforts. See also Michael R. Fein, *Paving the Way: New York Road Building and the American State, 1880–1956* (University of Kansas Press, 2008), 228; Christopher Wells, *Car Country: An Environmental History* (University of Washington Press, 2012), 273–74; and Lewis, *Divided Highways*, 114–21.

5. On earlier approaches to highway financing, see Roger Biles, "Expressways Before the Interstates: The Case of Detroit, 1945–1956," *Journal of Urban History* 40, no. 5 (2014): 843–54; and Christopher Wells, "Fueling the Boom: Taxes, Invisibility, and the Growth of the American Highway Infrastructure, 1919–1956," in "Oil in American History," special issue, *Journal of American History* 99, no. 1 (2012): 72–81. On the rise of bonded debt and its social consequences in this period, see Gail Radford, "From Municipal Socialism to Public Authorities: Institutional Factors in the Shaping of American Public Enterprise," *Journal of American History* 90, no. 3 (2003): 863–90; and Destin Jenkins, *The Bonds of Inequality: Debt and the Making of the American City* (University of Chicago Press, 2021). In quantitative terms, the postwar period saw "the total amount of state and local bonded debt nationwide grew sixfold, from $19 billion in 1948 to $124 billion in 1968," and funded "roads and bridges, sewers and waterworks, parks, and public housing"; see Michael Glass, "Schooling Suburbia: The Politics of School Finance in Postwar Long Island" (PhD diss., Princeton University, 2020), 23–24.

6. On the particularities of bonds and the manner in which they mediate the public-private relationship, see Jenkins, *Bonds of Inequality*, 1–9, 23–24. See also Eric H. Monkkonen, *The Local State: Public Money and American Cities* (Stanford University Press, 1995); and Alberta M. Sbragia, *Debt Wish: Entrepreneurial Cities, U.S. Federalism, and Economic Development* (University of Pittsburgh Press, 1996). On "associational" and "out of sight" governance, see Brian Balogh, *The Associational State: American Governance in the Twentieth Century* (University of Pennsylvania Press, 2015); and Brian Balogh, *A Government Out of Sight: The Mystery of National Authority in Nineteenth-Century America* (Cambridge University Press, 2009).

7. James Sparrow, *Warfare State: World War II Americans and the Age of Big Government* (Oxford University Press, 2011), 243; Robert A. Taft, "Progress Through Liberty: Speech to the Ohio Republican Rally," September 17, 1952, Springfield, OH, reprinted in Clarence E. Wunderlin Jr., ed., *The Papers of Robert A. Taft*, vol. 4, *1949–1953* (Kent State University Press, 2006), 421–24. Eisenhower is quoted in Kevin M. Kruse, *One Nation Under God: How Corporate America Invented Christian America* (Basic Books, 2015), 59–60.

8. Eisenhower is quoted in Gary Gerstle, *Liberty and Coercion: The Paradox of American Government from the Founding to the Present* (Princeton University Press, 2015), 266.

9. Several studies mention how Eisenhower's political views led him to support the basic concept of the interstate system, but presidential preferences are hardly a satisfying explanation

of legislative development. See Cohen, *A Consumer's Republic*, 118–19; Kruse, *One Nation Under God*, 85–87; and Lawrence Glickman, *Free Enterprise: An American History* (Yale University Press, 2019), 149–51.

10. On indebtedness to private interests and profit-seeking creditors, see Jenkins, *Bonds of Inequality*, 23–24. On the possibilities of more direct financial methods of state building provided by excise taxes, see Julian Zelizer, *Taxing America: Wilbur D. Mills, Congress, and the State* (Cambridge University Press, 1998), 55–61.

11. H. S. Fairbank, chief of Information Division, Bureau of Public Roads, Address at the 22nd Annual Convention of the American Association of State Highway Officials, San Francisco, CA, 1936, box 1, entry 12A, record group 30, Office Files and Correspondence Concerning Particular Bureau Activities, Records of the Bureau of Public Roads, National Archives and Records Administration, College Park, MD (hereafter cited as NARA), 3–4, 7.

12. The bureaucratic plans cohered in two key reports from the Bureau of Public Roads that prefigured the 1944 act: *Toll Roads and Free Roads* (1939) and *Interregional Highways* (1944). See also Wells, *Car Country*, 269–74.

13. Gallup Poll #1938-0113: War Debts / Automobiles / Roosevelt, Gallup Organization (Roper Center for Public Opinion Research, Cornell University, 1938), dataset, https://doi.org/10.25940/ROPER-31087097; Bureau of Public Roads, *Toll Roads and Free Roads*, 1–5, 13–15; Wells, *Car Country*, 196; Gallup Poll #1952-0492: Politics / Armed Forces, Gallup Organization (Roper Center for Public Opinion Research, Cornell University, 1952), dataset, https://doi.org/10.25940/ROPER-31087476; Dwight Eisenhower, "Annual Budget Message to the Congress: Fiscal Year 1955," online by Gerhard Peters and John T. Woolley, American Presidency Project, https://www.presidency.ucsb.edu/documents/annual-budget-message-the-congress-fiscal-year-1955.

14. The Eisenhower administration's aversion to debt and its concerns with efficient governmental management came on full display in Eisenhower's first State of the Union Address and continued to shape the executive branch's rhetoric and ambitions. See Dwight D. Eisenhower, "Annual Message to Congress on the State of the Union," February 2, 1953, Peters and Woolley, American Presidency Project, https://www.presidency.ucsb.edu/documents/annual-message-the-congress-the-state-the-union-16; Kim Phillips-Fein, *Invisible Hands: The Making of the Conservative Movement from the New Deal to Reagan* (W. W. Norton, 2009), 56, 166–67; Kruse, *One Nation Under God*, 87; and Glickman, *Free Enterprise*, 149–51. The $27 billion projection came from the President's Advisory Committee on a National Highway Program; see *A Ten-Year National Highway Program* (US Government Printing Office, 1955).

15. "Address of Vice President Richard Nixon to the National Governors Association Conference, Lake George, New York, July 12, 1954," box 1, entry 63, Records Relating to National Highway and Defense Highway Programs, 1940–55, Records of the Bureau of Public Roads, Record Group 30, NARA, 1.

16. Chamber of Commerce of the United States of America, *Socialism in America: A Study* (Chamber of Commerce of the United States of America, 1950), 2; "Address of Vice President Richard Nixon," 3–4. See also Matthew Lassiter, "Political History Beyond the Red-Blue Divide," *Journal of American History* 98, no. 3 (2011): 760–64; and Glickman, *Free Enterprise*, 15–16.

17. "Address of Vice President Richard Nixon," 2–3. For a conceptual understanding of planned economies in this period, see James C. Scott, *Seeing Like a State: How Certain Schemes to Improve the Human Condition Have Failed* (Yale University Press, 1998). For perspectives more closely tethered to United States history, see David Ekbladh, *The Great American Mission: Modernization and the Construction of an American World Order* (Princeton University Press,

2009); and Amy C. Offner, *Sorting Out the Mixed Economy: The Rise and Fall of Welfare and Developmental States in the Americas* (Princeton University Press, 2019).

18. "Address of Vice President Richard Nixon," 2, 5–6.

19. Taft, "Progress Through Liberty," 421–24. On the post–World War II adaptation of the New Deal state building framework see Jason Scott Smith, *Building New Deal Liberalism: The Political Economy of Public Works, 1933–1956* (Cambridge University Press, 2006), 232–57; and Jonathan Levy, *Ages of American Capitalism: A History of the United States* (Random House, 2021), 462–87. On the slogan "free enterprise" in conservative attacks on the New Deal, see Glickman, *Free Enterprise*, 144–51, 157–66.

20. "Address of Vice President Richard Nixon," 7.

21. James C. Hagerty, "Press Release, September 7th, 1954," box 1, entry 63, Records Relating to the National Highway and Defense Highway Programs, 1940–55, Records of the Bureau of Public Roads, Record Group 30, NARA; C. D. Curtiss to Francis Case, December 17, 1954, box 1, entry 6-C, General Correspondence and Related Records, 1951–55, Records of the Bureau of Public Roads, Record Group 30, NARA; *Congressional Record*, 84 Cong., 1st sess., July 27, 1955, 11691. See also Jean Edward Smith, *Lucius D. Clay: An American Life* (Henry Holt, 1990), 617–21. The composition of the Clay Committee reflected a broader trend in the Eisenhower administration, which was dominated by men with overt business interests; see Kruse, *One Nation Under God*, 84–85.

22. Oscar H. Nielson to Henry S. Bridges, October 13, 1954, box 1, entry 6-C, General Correspondence and Related Records, 1951–55, Records of the Bureau of Public Roads, Record Group 30, NARA.

23. Francis V. du Pont to Thomas Dewey, January 23, 1953, file 26, box 4, Political Correspondence, Papers of Francis V. du Pont, Hagley Museum and Library, Wilmington, DE (hereafter cited as Hagley); Francis V. du Pont to Thomas Dewey, February 13, 1953, file 26, box 4, Political Correspondence, Papers of Francis V. du Pont, Hagley. See also Rose and Mohl, *Interstate*, 70. On the du Pont family's business and political endeavors, see Phillips-Fein, *Invisible Hands*.

24. Francis V. du Pont to Harold E. Stassen, November 24, 1952, file 26, box 4, Political Correspondence, Francis V. du Pont Papers, Hagley.

25. Irénée du Pont to Francis V. du Pont, September 24, 1951, file 23, box 4, General Correspondence, Francis V. du Pont Papers, Hagley.

26. President's Advisory Committee on a National Highway Program, *Hearings, October 7–8, 1954*, box 1, entry 63, Records Relating to National Highway and Defense Highway Programs, 1940–55, Record Group 30, NARA.

27. President's Advisory Committee on a National Highway Program, *A Ten-Year National Highway Program*, xiii, 23–25, 29.

28. President's Advisory Committee on a National Highway Program, xiii, 23–25, 29.

29. William H. Stringer, "Democrats Criticize Plan to Finance Highways," *Christian Science Monitor*, February 23, 1955, 6.

30. Joseph A. Loftus, "Ike Submits Road Plan 'Vetoed' by Democrats," *Atlanta Constitution* (*New York Times News Service*, special to the *Atlanta Constitution*), February 23, 1955, 1, 13.

31. Joseph A. Loftus, "President Fails to Impress Democrats with Road Plan," *New York Times*, February 22, 1955, 1, 10.

32. Robert J. Donovan, "Highways Message Today; Democrats Get a Preview," *New York Herald Tribune*, February 22, 1955, 1, 15; "Which Road Is Right?," *St. Louis Post-Dispatch*, February 23, 1955, 2B.

33. "Democrats Not Impressed by Ike Road Plans," *Minneapolis Morning Tribune*, February 22, 1955, 3; Donovan, "Highways Message Today," 1, 15.

34. Loftus, "President Fails to Impress Democrats with Road Plan," 1, 10; "A Controversial Project," *Boston Globe*, February 23, 1955, 12. On Byrd's segregationist politics, see Deborah Archer, *Dividing Lines: How Transportation Infrastructure Reinforces Racial Inequality* (W. W. Norton, 2025), 21.

35. Loftus, "President Fails to Impress Democrats with Road Plan," 1, 10.

36. "The President's Highway Program," *Philadelphia Inquirer*, February 23, 1955, 20.

37. "Road Plan Outlined to Congress," *Los Angeles Times*, February 22, 1955, 32; Robert J. Donovan, "101-Billion Road Plan Offered by Eisenhower," *New York Herald Tribune*, February 23, 1955, 1, 27.

38. "Ike's Road Plan Foes Unyielding," *Atlanta Constitution*, February 22, 1955, 1.

39. Donovan, "Highways Message Today," 1, 15; See also Wendy L. Wall, *Inventing the "American Way": The Politics of Consensus from the New Deal to the Civil Rights Movement* (Oxford University Press, 2008).

40. *St. Louis Post-Dispatch*, "Which Road Is Right?," 2B.

41. "Democrats Not Impressed by Ike Road Plans," *Minneapolis Morning Tribune*, February 22, 1955.

42. William Knighton Jr., "Eisenhower Asks Passage of Roads Plan," *Baltimore Sun*, February 23, 1955, 1, 6; Stringer, "Democrats Criticize Plan to Finance Highways," 6.

43. Francis V. du Pont, untitled newspaper clipping from *GOP News*, January 17, 1950, box 11, Scrapbook 1949–61, Francis V. du Pont Papers, Hagley.

44. "F. V. du Pont's Radio Speech, Georgetown, Delaware, August 25, 1952," box 5, folder 39, Speeches of Francis V. du Pont, Hagley.

45. Du Pont, untitled newspaper clipping from *GOP News*.

46. House Committee on Public Works, *National Highway Program: Hearings Before the Committee on Public Works, Part 1*, 84 Cong., 1st Sess., April 19, 1955, 96.

47. Senate Committee on Public Works, *National Highway Program: Hearings Before a Subcommittee of the Committee on Public Works*, 84 Cong., 1st Sess. March 29, 1955, 655.

48. Senate Committee on Public Works, *National Highway Program*, March 11, 1955, 395–96.

49. House Committee on Public Works, *National Highway Program, Part 1*, April 18, 1955, 23.

50. House Committee on Public Works, *National Highway Program, Part 1*, April 18, 1955, 23–25.

51. Senate Committee on Public Works, *National Highway Program*, March 22, 1955, 539–41.

52. House Committee on Public Works, *National Highway Program, Part 1*, May 16, 1955, 579–81.

53. House Committee on Public Works, *National Highway Program, Part 1*, May 16, 1955, 600.

54. Senator Edward Martin, quoted in *Congressional Record*, 84 Cong., 1st sess., May 25, 1955, 6933.

55. House Committee on Public Works, *National Highway Program, Part 1*, May 16, 1955, 579, 600–601; Senate Committee on Public Works, *National Highway Program*, March 22, 1955, 541. On debt in American consumer culture in this period, see Louis Hyman, *Debtor Nation: The History of America in Red Ink* (Princeton University Press, 2011), esp. chap. 5, "Postwar Consumer Credit."

56. House Committee on Public Works, *National Highway Program, Part 1*, May 16, 1955, 592, 596.

57. Senate Committee on Public Works, *National Highway Program*, March 22, 1955, 541.

58. Senate Committee on Public Works, *National Highway Program*, March 22, 1955, 541–42,

551–52. Humphrey did not elaborate on the terminology that he chose, but he may have been thinking of so-called "moral obligation bonds," which Gail Radford describes as a type of mid-century "financial instrument [that] was essentially a revenue bond secured not only by income from the authority that issued it, but also by some kind of vaguely worded promise." But the fact that Humphrey did not specify this suggests he knew Democrats would find even more cause for concern with such an arrangement, given the terms of the debate. See Gail Radford, *The Rise of Public Authority: Statebuilding and Economic Development in Twentieth-Century America* (University of Chicago Press, 2013), 147.

59. Senate Committee on Public Works, *National Highway Program*, March 22, 1955, 542; House of Representatives, *National Highway Program: Hearings Before the Committee on Public Works, Part 2*, 84 Cong., 1st sess., July 12, 1955, 1193.

60. Ronald L. Heinemann, *Harry Byrd of Virginia* (University of Virginia Press, 1996), 251, 259, 312–16, 355; Senate Committee on Public Works, *National Highway Program*, March 18, 1955, 497–98, 500; Rose and Mohl, *Interstate*, 78.

61. Senate Committee on Public Works, *National Highway Program*, March 18, 1955, 497–500; *Congressional Record*, 84 Cong., 1st sess., January 18, 1955, 473.

62. Senate Committee on Public Works, *National Highway Program*, March 18, 1955, 497–500, 520; *Congressional Record*, 84 Cong., 1st sess., May 25, 1955, 7033.

63. *Congressional Record*, 84 Cong., 1st sess., July 27, 1955, 11691-4.

64. *Congressional Record*, 84 Cong., 1st sess., July 27, 1955, 11694-5, 11717-8; Social Security Act of 1935, Pub. Law 74-271, 49 Stat. 620 (1935).

65. Senate Committee on Finance, *Highway Revenue Act: Hearings Before the Committee on Finance*, 84 Cong., 2nd sess., May 17, 1956, 66.

66. House Committee on Ways and Means, *Highway Revenue Act of 1956: Hearings Before the Committee on Ways and Means*, 84 Cong., 2nd sess., February 14, 1956, 25.

67. *Congressional Record*, 84 Cong., 2nd sess., April. 27, 1956, 7221-2; *Congressional Record*, 84 Congress, 2nd sess., June 26, 1956, 10969. On earmarking, excise-based tax-and-spend mechanisms, and "special funds," see Zelizer, *Taxing America*, 17–18, 147–78.

68. Ben Kelley, *The Pavers and the Paved: The Real Cost of America's Highway Program* (Donald Brown, 1971), 27; Heinemann, *Harry Byrd of Virginia*, 364.

69. Senate Committee on Finance, *Highway Revenue Act: Hearings Before the Committee on Finance*, 84 Cong., 2nd sess., May 17, 1956, 66; House Committee on Ways and Means, *Highway Revenue Act of 1956: Hearings Before the Committee on Ways and Means*, 84 Cong., 2nd sess., February 14, 1956, 25; *Congressional Record*, 84 Cong., 2nd sess., April. 27, 1956, 7221-2; *Congressional Record*, 84 Cong., 2nd sess., June 26, 1956, 10969. *The New York Times* blamed the administration bill's defeat on the "anxious and turbulent" character of its executive branch backers; see "The News of the Week in Review," *New York Times*, July 31, 1955.

70. Pub. Law 84-627, 70 Stat. 374.

71. "Hike Gas, Tire Tax at Midnight," *Newsday*, June 30, 1956, 2.

72. Harry Stanton, "Fill Your Gas Tank, Buy That Tire Today Before Tax Goes Up," *Boston Globe*, June 30, 1956, 1–2; Senate Committee on Public Works, *National Highway Program*, March 22, 1955, 539–41.

73. Federal-Aid Highway Act of 1956, Pub. Law 84-627, 70 Stat. 374 (1956).

74. Radford, *Rise of Public Authority*; Jenkins, *Bonds of Inequality*.

75. Excise taxes on gasoline and other automotive products, of course, signaled postwar faith in oil supplies and cheap petrodollars as engines of state building. See Timothy Mitchell,

Carbon Democracy: Political Power in the Age of Oil (Verso, 2011), 109–43. See also Balogh, *Government Out of Sight.*

76. Critiques proliferated in the 1960s and 1970s, with eye-catching titles. See A. Q. Mowbray, *Road to Ruin: A Critical View of the Federal Highway Program* (Lippincott, 1969); Helen Leavitt, *Superhighway—Superhoax* (Doubleday, 1970); and Kelley, *The Pavers and the Paved.* See also Wells, *Car Country*, 275.

77. Gallup Poll #1956-0563: Politics / Middle East / Presidential Election, Gallup Organization (Roper Center for Public Opinion Research, Cornell University, 1956), dataset, https://doi.org/10.25940/ROPER-31087547; Gallup Poll #1938-0113: War Debts / Automobiles / Roosevelt, Gallup Organization (Roper Center for Public Opinion Research, Cornell University, 1938), dataset, https://doi.org/10.25940/ROPER-31087097.

78. Wells, *Car Country*, 274–77. For an overview of "policy feedback," see Julian Zelizer, "The Unexpected Endurance of the New Deal Order: Liberalism in the Age of Reagan," in *Beyond the New Deal Order: U.S. Politics from the Great Depression to the Great Recession*, ed. Gary Gerstle, Nelson Lichtenstein, and Alice O'Connor (University of Pennsylvania Press, 2019), 77. See also Paul Pierson, *Politics in Time: History, Institutions, and Social Analysis* (Princeton University Press, 2004), esp. chap. 1, "Positive Feedback and Path Dependence."

79. Francesca Russello Ammon, *Bulldozer: Demolition and the Clearance of the Postwar Landscape* (Yale University Press, 2016), 5, 182–218, 293; Adam Rome, *The Bulldozer in the Countryside: Suburban Sprawl and the Rise of American Environmentalism* (Cambridge University Press, 2001); Raymond Mohl, "Stop the Road: Freeway Revolts in American Cities," *Journal of American History* 30, no. 5 (2004): 674–706; and Michael R. Fein, "Highways and Livability Policy in the Post-Interstate Era, 1978–2013," *Journal of Urban History* 40, no. 5 (2014): 855–69.

80. Kelley, *The Pavers and the Paved*, 27.

81. Kelley's was one of several critical interpretations of the highway trust fund that emerged at the time. See also Mowbray, *Road to Ruin*, 18–43; and Leavitt, *Superhighway—Superhoax*, 230–33.

82. Kelley, *The Pavers and the Paved*, 27.

83. Ben Kelley and Richard Hébert, "Priorities or Trust Funds?," *The Nation*, April 19, 1971, 500; Mohl, "The Interstates and the Cities," 37; Senate Committee on Public Works, *National Highway Program*, March 18, 1955, 497.

84. Linda Gordon, *Pitied but Not Entitled: Single Mothers and the History of Welfare, 1890–1935* (Harvard University Press, 1994), 3–4. See also Theda Skocpol, *Protecting Soldiers and Mothers: The Political Origins of Social Policy in the United States* (Belknap Press of Harvard University Press, 1992); Robyn Muncy, *Creating a Female Dominion in American Reform, 1890–1935* (Oxford University Press, 1991); Suzanne Mettler, *Dividing Citizens: Gender and Federalism in New Deal Public Policy* (Cornell University Press, 1998); Felicia Kornbluh, *The Battle for Welfare Rights: Politics and Poverty in Modern America* (University of Pennsylvania Press, 2007); and Karen M. Tani, *States of Dependency: Welfare, Rights, and American Governance, 1935–1972* (Cambridge University Press, 2016).

Chapter Three

1. Phillip B. Yeager, "Your State May Block New Roads," *The Nation's Business*, September 1956, 96–97, 100–103. See also Frank C. Moore, "The State's Responsibility for State-Local Action," in *Financing Highways, Symposium, Princeton, NJ, November 8–9, 1956* (Tax Institute, 1957), 74.

2. H. S. Fairbank, "Objects and Methods of the State-Wide Highway Planning Surveys,"

American Association of State Highway Officials Annual Convention, San Francisco, California, December 8, 1936, box 1, entry 12-A, Office Files and Correspondence Concerning Particular Bureau Activities, Records of the Bureau of Public Roads, Record Group 30, 11.

3. Homer E. Scace, foreword to *Financing Highways*, iii.

4. Edward A. Lutz, "Effect of New Federal Aid Highway Act on Federal-State Relations," in *Financing Highways*, 57.

5. Moore, "The State's Responsibility for State-Local Action," 71.

6. Cotten Seiler, *Republic of Drivers: A Cultural History of Automobility in America* (University of Chicago Press, 2008), 78–79.

7. James C. Scott, *Domination and the Arts of Resistance: Hidden Transcripts* (Yale University Press, 1990), 18, 49.

8. George Romney, "The Motor Vehicle and the Highway: Some Historical Implications," in *Highways in Our National Life*, ed. Jean Labatut and Wheaton J. Lane (Princeton University Press, 1950), 220–21.

9. Seiler, *Republic of Drivers*, 79–80.

10. Ford Motor Company, *Freedom of the American Road* (Ford Motor Company, 1956), 1.

11. Bernard DeVoto, "The American Road," in Ford Motor Company, *Freedom of the American Road*, 8.

12. Ford Motor Company, 7; Frederick Jackson Turner, "The Significance of the Frontier in American History," American Historical Association conference, Chicago, July 12, 1893.

13. Deborah Archer, *Dividing Lines: How Transportation Infrastructure Reinforces Racial Inequality* (W. W. Norton, 2025), 7; Eric Avila, *The Folklore of the Freeway: Race and Revolt in the Modernist City* (University of Minnesota Press, 2014), 3, 40–41, Nathan D. B. Connolly, *A World More Concrete: Real Estate and the Remaking of Jim Crow South Florida* (University of Chicago Press, 2014), 8.

14. Bureau of Public Roads, *Toll Roads and Free Roads* (US Government Printing Office, 1939), 99; Bureau of Public Roads, *Interregional Highways* (US Government Printing Office, 1944), 80, 87.

15. The ARBA is quoted in Helen Leavitt, *Superhighway—Superhoax* (Doubleday, 1970), 122.

16. Johnson is quoted in Archer, *Dividing Lines*, 11. See also Connolly, *A World More Concrete*, 8, 261.

17. Seiler, *Republic of Drivers*, 84; "Construction: March of the Monsters," *Time*, June 24, 1957, https://time.com/archive/6887873/construction-march-of-the-monsters/.

18. Seiler, *Republic of Drivers*, 78–79.

19. President's Advisory Committee on a National Highway Program, *A Ten-Year National Highway Program* (US Government Printing Office, 1955), 4.

20. Alice Kessler-Harris, *In Pursuit of Equity: Women, Men, and the Quest for Economic Citizenship in 20th-Century America* (Oxford University Press, 2001), 211.

21. Kessler-Harris, *In Pursuit of Equity*, 205.

22. Association of MIT Alumnae, "Reports on Status of Women at MIT," accessed January 14, 2026, https://amita.alumgroup.mit.edu/s/1314/bp19/interior.aspx?sid=1314&gid=20&pgid=56230; Alex McGee, "Well-Behaved Women Should Make History," a review of MIT class records between 1888 and 1893, MIT Library, accessed August 1, 2025, https://libraries.mit.edu/news/well-behaved-women/30691/; "Bridge to the Bicentennial," April 17, 2025, *Every Day Matters* (blog), Rensselaer Polytechnic Institute, https://everydaymatters.rpi.edu/first-women-graduates/; Tom

Lewis, *Divided Highways: Building The Interstate Highways, Transforming American Life* (1997; Cornell University Press, 2013), 131–32.

23. See, for example, economic figures compiled by the Tax Institute in *Financing Highways*.

24. Linda Gordon, *Pitied but Not Entitled: Single Mothers and the History of Welfare* (Harvard University Press, 1994), 215.

25. Richard Whalen, "The American Highway: Do We Know Where We're Going?" *Saturday Evening Post*, December 14, 1968; Betty Friedan, *The Feminine Mystique* (1963; W. W. Norton, 2001), 63.

26. Seiler, *Republic of Drivers*, 87–88.

27. Roland Marchand, "Visions of Classlessness, Quests for Domination: American Popular Culture, 1945–1960," in *Reshaping America: Society and Institutions, 1945–1960*, ed. Robert Bremner and Gary Reichard (Ohio State University Press, 1982), 172.

28. Quoted in Seiler, *Republic of Drivers*, 87.

29. Seiler, 87–88.

30. Jack Kerouac, *On the Road* (Viking, 1957).

31. "He wanted a Fiat Spider," Fiat advertisement, original publication unknown, clipping in author's possession.

32. "The engine is idling," Plymouth advertisement, original publication unknown, clipping in author's possession.

33. Victor Eichhorn, "Relocation Assistance," *Proceedings of the American Association of State Highway Officials* (American Association of State Highway Officials, 1968), 98.

34. C. G. Prahl, "Random Uncensored Observations of a New Highway Director," *Proceedings of the American Association of State Highway Officials* (American Association of State Highway Officials, 1965), 1.

35. Ellis L. Armstrong, "Cohesion—Soils and People," remarks delivered to the American Society of Civil Engineers Research Conference, Boulder, CO, June 16, 1960, folder 47, box 6, Bureau of Public Roads, Papers of Francis V. du Pont, Hagley Museum and Library, Wilmington, DE (hereafter cited as Hagley), 9.

36. David Durenberger, "Remarks on Public Relations as Viewed from a Governor's Office," *Proceedings of the American Association of State Highway Officials* (American Association of State Highway Officials, 1968), 146.

37. The process by which these engineers obtained policymaking power was complex and unfolded over several decades during the first half of the twentieth century. The key study here is Bruce Seely, *Building the American Highway System: Engineers as Policy Makers* (Temple University Press, 1987).

38. Bureau of Public Roads *General Location of the National System of Interstate Highways, Including All Additional Routes at Urban Areas Designated in September 1955* (US Government Printing Office, 1955).

39. Francesca Russello Ammon, *Bulldozer: Demolition and the Clearance of the Postwar Landscape* (Yale University Press, 2016), 197.

40. Bureau of Public Roads, *America's Lifelines: Federal Aid for Highways* (US Government Printing Office, 1962), 9.

41. Ammon, *Bulldozer*, 190. See also Lewis Mumford, "The Highway and the City," *Architectural Record*, April 1958, 179–186.

42. Richard F. Weingroff, "The Greatest Decade, 1956–1966," part 2, "The Battle of Its Life," Highway History, Federal Highway Administration, updated June 27, 2017, https://www.fhwa

.dot.gov/infrastructure/50interstate2.cfm; Arnold H. Vollmer and Robert L. Rotner, "The Role of the Architect in Highway Planning," *Proceedings of the American Association of State Highway Officials* (American Association of State Highway Officials, 1966), 65–66.

43. National Aeronautics and Space Administration, "Chronological History, Fiscal Year 1959–1979, Budget Submissions," accessed February 9, 2026, https://www.nasa.gov/wp-content/uploads/2023/12/o45128943-1959-1979.pdf; Ben Kelley, *The Pavers and the Paved: The Real Cost of America's Highway Program* (Donald Brown, 1971), 3, 7.

44. Weingroff, "The Greatest Decade."

45. Kelley, *The Pavers and the Paved*, 3–7; Michael R. Fein, *Paving the Way: New York Road Building and the American State, 1880–1956* (University of Kansas Press, 2008).

46. Bureau of Public Roads, *America's Lifelines* (1962), 4–5.

47. Seiler, *Republic of Drivers*, 94–96.

48. Bureau of Public Roads, *America's Lifeline* (1962), 8, 11. On Cold War statist fears on the right, see Glickman, *Free Enterprise*. The literature on policy battles waged at the meeting points of federalism's layers is truly vast. A useful overview is Gary Gerstle, *Liberty and Coercion: The Paradox of American Government from the Founding to the Present* (Princeton University Press, 2015).

49. Bureau of Public Roads, *America's Lifelines* (1962), 4, 6, 9.

50. Kelley, *The Pavers and the Paved*, 4; Bureau of Public Roads, *America's Lifelines: Federal Aid for Highways* (US Government Printing Office, 1966), 9–11.

51. Adam Rome, *The Bulldozer in the Countryside: Suburban Sprawl and the Rise of American Environmentalism* (Cambridge University Press, 2001), 120.

52. Ammon, *Bulldozer*, 85–89; Rome, *The Bulldozer in the Countryside*, 121.

53. Ammon, *Bulldozer*, 90–92.

54. Ammon, 90–92.

55. Ammon, 185–86, 194–95, 198.

56. For historical climate data see National Centers for Environmental Information, National Oceanic and Atmospheric Administration, https://www.ncdc.noaa.gov/cdo-web/; Ammon, *Bulldozer*, 185–86, 194–95, 198.

57. News of these astounding environmental modifications is cited in Kelley, *The Pavers and the Paved*, 13–14.

58. "Road Builders' Biggest Year," *Life*, December 12, 1949, 142.

59. "Investor's Guide," *Chicago Daily Tribune*, July 1, 1956.

60. Weingroff, "The Greatest Decade."

61. Lewis, *Divided Highways*, 144.

62. Richard F. Weingroff, "The Man Who Changed America, Part II," Highway History, Federal Highway Administration, accessed August 1, 2025, https://www.fhwa.dot.gov/publications/publicroads/03may/05.cfm.

63. Weingroff, "The Greatest Decade."

64. Karen Tani, *States of Dependency: Welfare, Rights, and American Governance, 1935–1972* (Cambridge University Press, 2016), 202.

65. Lutz, "Effect of New Federal Aid Highway Act," 63–64; Moore, "The State's Responsibility for State-Local Action," 76.

66. Lutz, "Effect of New Federal Aid Highway Act," 63–64; Moore, "The State's Responsibility for State-Local Action," 76.

67. Lutz, "Effect of New Federal Aid Highway Act," 64.

68. Richard F. Weingroff, "Essential to the National Interest," Highway History, Federal Highway Administration, accessed August 1, 2025, https://www.fhwa.dot.gov/publications/publicroads/06mar/07.cfm; Weingroff, "The Greatest Decade;" Ammon, *Bulldozer*, 183, 195–97.

69. Lewis, *Divided Highways*, 129–30; Weingroff, "Essential to the National Interest."

70. United States Department of Commerce press release, January 4, 1957, box 10, entry 12G, Office Files and Correspondence Concerning Particular Bureau Activities, Records of the Bureau of Public Roads, Record Group 30, NARA; Weingroff, "Essential to the National Interest."

71. United States Department of Commerce press release, January 4, 1957.

72. Weingroff, "The Greatest Decade."

73. Bureau of Public Roads, *America's Lifelines* (1962), 3–4.

74. Weingroff, "The Greatest Decade." See also Bureau of Public Roads, *America's Lifelines* (1962), 4–5.

75. Weingroff, "The Greatest Decade."

76. Scott, *Domination and the Arts of Resistance*, 45.

77. The concept of path dependence comes from political scientists and historical institutionalists. The key overview is Paul Pierson, *Politics in Time: History, Institutions, and Social Analysis* (Princeton University Press, 2004), esp. chap. 1, "Positive Feedback and Path Dependence."

78. David H. Stevens, "Public Acceptance—an Essential Part of a Highway Program," *Proceedings of the American Association of State Highway Officials* (American Association of State Highway Officials, 1956), 10.

79. Stevens, "Public Acceptance," 10–11.

80. Stevens, 11–12.

81. Stevens, 14.

82. The engineer is quoted in Ammon, *Bulldozer*, 210.

83. LeRoy Moser, "Negotiation Policy and Procedure," *Proceedings of the American Association of State Highway Officials* (American Association of State Highway Officials, 1956), 62–63.

84. Stevens, "Public Acceptance," 13–14.

Chapter Four

1. Lewis Mumford, "The Highway and the City," *Architectural Record*, April 1958, 179.

2. Mumford, "The Highway and the City," 181.

3. For examples of Lewis Mumford's work, see *The Story of Utopias* (Boni and Liveright, 1922), *The Culture of Cities* (Harcourt, Brace, 1938), and *City Development* (Harcourt, Brace, 1945). See also Aaron Sachs, "Lewis Mumford's Urbanism and the Problem of Environmental Modernity," *Environmental History* 21, no. 4 (2016): 638–59.

4. Mumford, "The Highway and the City," 181–82.

5. Francesca Russello Ammon, *Bulldozer: Demolition and the Clearance of the Postwar Landscape* (Yale University Press, 2016), 5, 293.

6. Ben Kelley, *The Pavers and the Paved: The Real Cost of America's Highway Program* (Donald Brown, 1971), 5.

7. Kelley, *The Pavers and the Paved*, 5.

8. David Hapgood, "The Highwaymen," *Washington Monthly*, March 1969, 77.

9. Ammon, *Bulldozer*, 183, 191, 204, 211. The unnamed official is quoted in Raymond Mohl, "Stop the Road: Freeway Revolts in American Cities," *Journal of Urban History* 30, no. 5 (2004):

680. See also Eric Avila, *The Folklore of the Freeway: Race and Revolt in the Modernist City* (University of Minnesota Press, 2014).

10. The freeway revolts only occasionally resulted in immediate material success. In the longer term, they may have contributed to broader and more successful cultural movements against "reigning ideals of progress" and "modernist" master planning. See Avila, *The Folklore of the Freeway*, 29.

11. J. C. Womack, "The Importance of Public Information to the Highway Program," *Proceedings of the American Association of State Highway Officials* (American Association of State Highway Officials, 1965), 175.

12. Quotations and other details appear in Richard F. Weingroff, "The Greatest Decade, 1956–1966," part 1, "Essential to the National Interest," Highway History, Federal Highway Administration, updated June 30, 2023, https://www.fhwa.dot.gov/infrastructure/50interstate.cfm.

13. Mark H. Rose and Raymond A. Mohl, *Interstate: Highway Politics and Policy Since 1939* (1979; University of Tennessee Press, 2012), 116; the resolution is quoted in Weingroff, "The Greatest Decade," part 1.

14. Chris Carlsson, "The Freeway Revolt," *FoundSF*, accessed January 15, 2026, https://www.foundsf.org/The_Freeway_Revolt. Pete Seeger's 1963 recording of "Little Boxes" reached number 70 on the Billboard charts in 1964; see Billboard Database's entry for the recording, accessed January 15, 2026, https://elpee.jp/single/Little%20Boxes/Pete%20Seeger/.

15. Jane Jacobs's organizing strategies made astute use of gender politics and leveraged "motherhood" as a source of legitimacy in community protection efforts. This led Jacobs's nemesis Robert Moses, seeking to bulldoze Washington Square Park, to claim bitterly that "there is nobody against this—nobody, nobody, nobody but a bunch of mothers." See Avila, *Folklore of the Freeway*, chap. 2, "Nobody but a Bunch of Mothers"; Moses is quoted at 64. Broadly, the demographic and political contours of the freeway revolts are examined in Mohl, "Stop the Road," 674–80. See also Raymond Mohl, "The Interstates and the Cities: The U.S. Department of Transportation and the Freeway Revolt, 1966–73," *Journal of Policy History* 20, no. 2 (2008), 193–226.

16. Rose and Mohl, *Interstate*, 116; A. Q. Mowbray, *Road to Ruin: A Critical View of the Federal Highway Program* (Lippincott, 1969), 97.

17. Mumford, "The Highway and the City," 186.

18. United States Department of Commerce, *Quarterly Report on the Federal-Aid Highway Program, June 30, 1960*, box 10, entry 12G, Office Files and Correspondence Concerning Particular Bureau Activities, Records of the Bureau of Public Roads, Record Group 30, National Archives and Records Administration, College Park, MD (hereafter cited as NARA).

19. Karl Detzer, "Our Great Big Highway Bungle," *Reader's Digest*, July 1960, 50.

20. Angus Burgin, *The Great Persuasion: Reinventing Free Markets Since the Depression* (Harvard University Press, 2012), 88–89; Kim Phillips-Fein, *Invisible Hands: The Making of the Conservative Movement from the New Deal to Reagan* (W. W. Norton, 2009), 40–42.

21. Friedrich Hayek, *The Road to Serfdom* (1944; University of Chicago Press, 1994); Phillips-Fein, *Invisible Hands*, 40–42; Burgin, *The Great Persuasion*, 88–89. See also Daniel Rodgers, *Age of Fracture* (Harvard University Press, 2012), and Lawrence Glickman, *Free Enterprise: An American History* (Yale University Press, 2019).

22. Detzer, "Our Great Big Highway Bungle," 45–46.

23. Byrd is quoted in Detzer, "Our Great Big Highway Bungle," 51. On Byrd's segregationist goals in connection with the interstate system, see Deborah Archer, *Dividing Lines: How Transportation Infrastructure Reinforces Racial Inequality* (W. W. Norton, 2025), 21.

24. Detzer, "Our Great Big Highway Bungle," 45–46.

25. Detzer, 46.

26. Brent Cebul and Mason B. Williams, "'Really and Truly a Partnership': The New Deal's Associational State and the Making of Postwar American Politics," in *Shaped by the State: Toward a New Political History of the Twentieth Century*, ed. Brent Cebul, Lily Geismer, and Mason B. Williams (University of Chicago Press, 2019), 96–122.

27. Detzer, "Our Great Big Highway Bungle," 45–47.

28. Detzer, 47–48. On interstate-borne segregation and suburbanization—and profit interests in both—in Georgia, see Kevin M. Kruse, *White Flight: Atlanta and the Making of Modern Conservatism* (Princeton University Press, 2005), 242–51.

29. Detzer, "Our Great Big Highway Bungle," 48.

30. Detzer, 49–50.

31. Jane Jacobs, *The Death and Life of Great American Cities* (1961; Vintage, 1992); Rachel Carson, *Silent Spring* (Houghton Mifflin, 1962); Ralph Nader, *Unsafe at Any Speed: The Designed-In Dangers of the American Automobile* (Grossman, 1965). On the car-defined American landscape, see Paul Sutter, *Driven Wild: How the Fight Against Automobiles Launched the Modern Wilderness Movement* (University of Washington Press, 2002), and Christopher Wells, *Car Country: An Environmental History* (University of Washington Press, 2012).

32. Richard Whalen, "The American Highway: Do We Know Where We're Going?" *Saturday Evening Post*, December 14, 1968.

33. Bureau of Public Roads, *America's Lifelines: Federal Aid for Highways* (US Government Printing Office, 1966), 7.

34. Ford Motor Company, *Freedom of the American Road* (Ford Motor Company, 1956).

35. Whalen, "The American Highway."

36. Whalen, "The American Highway."

37. Hapgood, "The Highwaymen," 11, 73.

38. Hapgood, 73. For more on Jacobs and community resistance to the project, see Avila, *The Folklore of the Freeway*, 57–65.

39. Mowbray, *Road to Ruin*, 34.

40. Whalen, "The American Highway."

41. Whalen, "The American Highway." Legal and historical research has affirmed what Whalen described. Officials routed highway development around white areas in Nashville, including the Belle Mead neighborhood, Vanderbilt University, and Centennial Park. They not only targeted the North Nashville community instead but appeared to go out of their way to wipe out the Black business district. See Archer, *Dividing Lines*, 153.

42. Whalen, "The American Highway."

43. Whalen, "The American Highway."

44. Richard F. Weingroff, "Addressing the Quiet Crisis: Origins of the National Environmental Policy Act of 1969," section 1, Highway History, Federal Highway Administration, accessed January 15, 2026, https://highways.dot.gov/highway-history/general-highway-history/addressing-quiet-crisis-origins-national-environmental; Whalen, "The American Highway"; David H. Stevens, "Public Acceptance—an Essential Part of a Highway Program," *Proceedings of the American Association of State Highway Officials* (American Association of State Highway Officials, 1956), 11.

45. The journalist is quoted in Ammon, *Bulldozer*, 210.

46. Whalen, "The American Highway."

47. Whalen, "The American Highway."

48. Whalen, "The American Highway."

49. Whalen, 60.

50. Whalen, "The American Highway."

51. Harold L. Plummer, "Highway Administrators Must Realize That Short Cuts and Advanced Techniques Are a 'Must' if Expanded Program Is to Succeed," *Proceedings of the American Association of State Highway Officials* (American Association of State Highway Officials, 1956), 4.

52. Plummer, "Highway Administrators Must Realize," 6.

53. James D. Dilts, "Pounding the Pavement," *Washington Monthly*, July 1969, 55.

54. Mowbray, *Road to Ruin*, 21.

55. Federal-Aid Highway Act of 1968, Pub. Law 90-495, 82 Stat. 815.

56. Mowbray, *Road to Ruin*, 20; Brian Francis, "Gasoline Excise Taxes, 1933–2000," Internal Revenue Service, accessed January 15, 2026, https://www.irs.gov/pub/irs-soi/00gastax.pdf.

57. Mowbray, *Road to Ruin*, 20–21.

58. Mowbray, 20–21; Hapgood, "The Highwaymen," 5. The Commission on Intergovernmental Relations is quoted in Bruce Seely, *Building the American Highway System: Engineers as Policy Makers* (Temple University Press, 1987), 227.

59. Mowbray, *Road to Ruin*, 35.

60. The Kansas official is quoted in Ammon, *Bulldozer*, 212.

61. Dilts, "Pounding the Pavement," 56.

62. Hapgood, "The Highwaymen," 80.

63. Robert G. Bartlett, "Remarks on Public Information," *Proceedings of the American Association of State Highway Officials* (American Association of State Highway Officials, 1968), 151.

64. David Durenberger, "Remarks on Public Relations as Viewed from a Governor's Office," *Proceedings of the American Association of State Highway Officials* (American Association of State Highway Officials, 1968), 147.

65. Kelley, *The Pavers and the Paved*, 16.

66. Robert Bartlett, "Highway Financing—Feast or Famine?" *Proceedings of the American Association of State Highway Officials* (American Association of State Highway Officials, 1968), 1.

67. Mowbray, *Road to Ruin*, 23–26.

68. Mowbray, 23–26.

69. Mowbray, 23–25, 29–30.

70. Ben Kelley and Richard Hébert, "Priorities or Trust Funds?," *The Nation*, April 19, 1971, 497–500.

71. The *Engineering News-Record* is quoted in Mowbray, *Road to Ruin*, 24.

72. Kelley, *The Pavers and the Paved*, 127–50.

73. Whalen, "The American Highway," 62; Kelley, *The Pavers and the Paved*, 127–28.

74. Arnold H. Vollmer and Robert L. Rotner, "The Role of the Architect in Highway Planning," *Proceedings of the American Association of State Highway Officials* (American Association of State Highway Officials, 1966), 66.

75. Holmes is quoted in Whalen, "The American Highway," 27.

76. Kelley, *The Pavers and the Paved*, 17.

77. Kelley, 17–18.

78. Henrik E. Stafseth, "Build, Baby, Build," *Proceedings of the American Association of State Highway Officials* (American Association of State Highway Officials, 1969), 1.

79. Harold C. King, "Highways in the Environment," *Proceedings of the American Association of State Highway Officials* (American Association of State Highway Officials, 1968), 352–53.

Chapter Five

1. Flem Otey's experience is recounted in Richard Whalen, "The American Highway: Do We Know Where We're Going?" *Saturday Evening Post*, December 14, 1968. Lizabeth Cohen tracks the development of interstate-highway-linked shopping centers and suburbs in *A Consumer's Republic: The Politics of Mass Consumption in Postwar America* (Vintage, 2003); see also Owen Gutfreund, *Twentieth-Century Sprawl: Highways and the Reshaping of the American Landscape* (Oxford University Press, 2004).

2. Whalen, "The American Highway."

3. The freeway revolts are enshrined in collective urban memories and live on in the nation's "folklore," as Eric Avila puts it. Inhabitants of America's cities recall and continue to experience the manner in which urban interstates cleaved their lives, a reality recounted by historians and legal scholars attuned to the racist nature of mid-twentieth-century urban state building. Yet the consequences of the freeway revolts in the development of the nation's law and governance have been underexplored. The legacy of interstate highway contestation lives on not just in folklore, but also in the law and its spatial configuration of the built environment. See Eric Avila, *The Folklore of the Freeway: Race and Revolt in the Modernist City* (Minneapolis: University of Minnesota Press, 2014).

4. In addition to Avila's *The Folklore of the Freeway*, a range of historical treatments explore the history—and cultural endurance—of the freeway revolts: See Francesca Russello Ammon, *Bulldozer: Demolition and the Clearance of the Postwar Landscape* (Yale University Press, 2016); Eric Avila, *Popular Culture in the Age of White Flight: Fear and Fantasy in Suburban Los Angeles* (University of California Press, 2004); Raymond Mohl, "Stop the Road: Freeway Revolts in American Cities," *Journal of Urban History* 30, no. 5 (2004): 674–706; and Raymond Mohl, "The Interstates and the Cities: The U.S. Department of Transportation and the Freeway Revolt, 1966–1973," *Journal of Policy History*, 20, no. 2 (2008): 193–226.

5. *Nashville I-40 Steering Committee v. Ellington*, 390 U.S. 921 (1968); "Flournoy A. Coles, Jr.," *Washington Post*, August 5, 1982; Theodore Cross and Robert Bruce Slater, "Michael Eric Dyson and the Vanderbilt Snub: Searching for a 'System Negro,'" *Journal of Blacks in Higher Education*, no. 24 (1999): 117–19; *Citizens to Preserve Overton Park, Inc. v. Volpe*, 401 U.S. 402 (1971). On *Overton Park*'s enduring significance in law and legal studies, see Reuel Schiller, "Enlarging the Administrative Polity: Administrative Law and the Changing Definition of Pluralism, 1945–1970," *Vanderbilt Law Review* 53, no. 5, (2000): 1389–454; Peter L. Strauss, ed., *Administrative Law Stories* (Foundation Press, 2006). Mid-twentieth-century American liberalism, of course, was replete with competing impulses among power wielders. Some sought a more egalitarian infrastructural future—generally in vain. The key treatment of this subject is Lizabeth Cohen, *Saving America's Cities: Ed Logue and the Struggle to Renew Urban America in the Suburban Age* (Farrar, Straus and Giroux, 2019). Elements of this story are also central to Kimberly Phillips-Fein, *Fear City: New York's Fiscal Crisis and the Rise of Austerity Politics* (MacMillan, 2018). Phillips-Fein tracks the economic history that worked in concert with the legal shifts under examination here to eviscerate municipal services, produce social disparities, and reshape the course of American governance. Meanwhile, Brent Cebul and Mason B. Williams have made visible the underlying structures of liberal governance that enabled infrastructure to be such a pliable tool, capable

of giving egalitarian reformers optimism yet also capable of etching disparity into the mid-twentieth-century American landscape. They call on scholars to "consider not just the structure of the New Deal state, but the intended and unintended uses to which communities, interest groups, and citizens have directed . . . the tools of New Deal federalism." See Cebul and Williams, " 'Really and Truly a Partnership': The New Deal's Associational State and the Making of Postwar American Politics," in *Shaped by the State: Toward a New Political History of the Twentieth Century*, ed. Brent Cebul, Lily Geismer, and Mason B. Williams (University of Chicago Press, 2019), 115. In brief, the interstate highway system was a powerful technology of governance at the center of a complex governmental context marked by competing interests and competing levels of federalism. That the interstate system had the effects it did—concretizing racial and class hierarchies despite pushback and countervailing ideas—is indicative of how federal administrators won the power struggle in the courts.

6. Legal scholars have commented on *Overton Park*. See William A. Thomas, "The Road to Overton Park: Parklands Statutes in Federal Highway Legislation," *Tennessee Law Review* 39, no. 3 (1972): 433–58; Peter L. Strauss, "Revisiting *Overton Park*: Political and Judicial Controls over Administrative Actions Affecting the Community," *UCLA Law Review* 39, no. 5 (1992): 1251–330; Daniel A. Farber, "Saving Overton Park: A Comment on Environmental Values," *University of Pennsylvania Law Review* 146, no. 5 (1998): 1671–86; and Tannera George Gibson, "Not in My Neighborhood: Memphis and the Battle to Preserve Overton Park," *University of Memphis Law Review* 41, no. 4 (2011): 725–44. Few historians have centered both grassroots social movements *and* ensuing legal developments; see Raymond A. Mohl, "Citizen Activism and Freeway Revolts in Memphis and Nashville: The Road to Litigation," *Journal of Urban History* 40, no. 5 (2014): 870–93. As Mohl's title indicates, among historians much of the emphasis has rested on social mobilization leading up to litigation, rather than on litigation itself—and its consequences.

7. Whalen, "The American Highway."

8. Ammon, *Bulldozer*, 183; Nathan D. B. Connolly, *A World More Concrete: Real Estate and the Remaking of Jim Crow South Florida* (University of Chicago Press, 2014), 8, 183; Deborah Archer, *Dividing Lines: How Transportation Infrastructure Reinforces Racial Inequality* (W. W. Norton, 2025), 24.

9. While much of the focus here rests on federal administrators, this was a key period of expanding subnational administrative and associational governance, as well. Important studies of local and state activity in this regard include Cebul and Williams, " 'Really and Truly a Partnership' "; Paul Sabin, *Public Citizens: The Attack on Big Government and the Remaking of American Liberalism* (Yale University Press, 2021); and Karen Tani, *States of Dependency: Welfare, Rights, and American Governance, 1935–1972* (Cambridge University Press, 2016).

10. Reuel Schiller, "The Era of Deference: Courts, Expertise, and the Emergence of New Deal Administrative Law," *Michigan Law Review* 106, no. 3 (2007): 404–6.

11. *Addison v. Holly Hill Fruit Farms*, 322 U.S. 607 (1944); Administrative Procedure Act of 1944, Pub. Law 79-404, 60 Stat. 237; Schiller, "Enlarging the Administrative Polity," 1417–19.

12. Louis L. Jaffe, *Judicial Control of Administrative Action* (Little, Brown, 1965), 323; Anne Kornhauser, *Debating the American State: Liberal Anxieties and the New Leviathan, 1930–1970* (University of Pennsylvania Press, 2015), 56; "Louis L. Jaffe, Noted Legal Scholar," *New York Times*, December 15, 1996.

13. Philip Selznick, *The TVA and the Grassroots* (1949; Harper and Row, 1966); Kornhauser, *Debating the American State*, 59, 82–89.

14. Robert Caro, *The Power Broker: Robert Moses and the Fall of New York* (Knopf, 1974).

15. On "sympathetic critics" of the administrative state in this period, see Kornhauser, *Debating the American State*. On conservative critics in this period, see Kim Phillips-Fein, *Invisible Hands: The Making of the Conservative Movement from the New Deal to Reagan* (W. W. Norton, 2009); Lawrence Glickman, *Free Enterprise: An American History* (Yale University Press, 2019), 149–51. On administrative oversight, see Daniel R. Ernst, *Tocqueville's Nightmare: The Administrative State Emerges in America, 1900–1940* (Oxford University Press, 2014); and Schiller, "Enlarging the Administrative Polity."

16. I determined these figures with the advanced search function in the electronic database Westlaw. Setting date ranges limited by the Federal-Aid Highway Acts of 1944 and 1956, I searched records of state and federal court cases for the terms "interstate highway," "interstate highway system," "federal highway," and "federal-aid highway act" and cross-referenced the results. However, it is important to note that the findings from Westlaw are almost certainly incomplete. Westlaw contains only reported decisions, but not all litigation is reported; settlement, for instance, might leave a state or federal court nothing to report. States also often do not report litigation from courts of first instance (trial courts), though I found many such cases reported nonetheless. Furthermore, not all states have consistently maintained intermediate courts; Alaska, for instance, did not have an appellate court until 1980, meaning that any interstate litigation there might not be reflected in Westlaw if it was not reported at the trial court level and did not reach the Alaska Supreme Court. We are left to consider how likely it is that additional trial court records would alter the temporal change reflected in the Westlaw data I collected—the key insight being that litigation, even if underreported, increased after 1956 when the interstates became physically real. On the challenges of underreported court data, see Tani, *States of Dependency*, 125–28.

17. Federal-Aid Highway Act of 1956, Pub. Law 84-627, 70 Stat. 374.

18. *United States v. Certain Parcels of Land*, SD Ill., 209 F. Supp. 483 (1962); *Department of Public Works & Buildings v. Ells*, Peoria County District Court, 23 Ill. 2d 619 (1962).

19. Mohl, "The Interstates and the Cities," 198; Joseph S. Clark, "Cities Revolt Against the Expressway," reprinted in *Congressional Record—Senate*, 89th Cong., 2d sess., 6 April 1966.

20. Federal-Aid Highway Act of 1966, Pub. Law 89-574, 80 Stat. 766.

21. Federal-Aid Highway Act of 1966, Pub. Law 89-574, 80 Stat. 766; Department of Transportation Act of 1966, Pub. Law 89-670 80, Stat. 931; Federal-Aid Highway Act of 1968, Pub. Law 90-495, 82 Stat. 815. On the term "Parklands Statutes," see Thomas, "The Road to Overton Park."

22. Boyd's quotes and other details here appear in Mohl, "The Interstates and the Cities," 200–201.

23. Mohl, "The Interstates and the Cities," 208–10.

24. Turner is quoted in Mohl, "The Interstates and the Cities," 201.

25. US Congress, *Hearings Before the Subcommittee on Roads of the Committee on Public Works, United States Senate, 90th Congress, First and Second Sessions, 1968* (US Government Printing Office, 1968).

26. US Congress, *Hearings Before the Subcommittee on Roads*, 249; Ammon, *Bulldozer*, 217.

27. US Congress, *Hearings Before the Subcommittee on Roads*, 249, 256, 488, 533.

28. *Nashville I-40 Steering Committee v. Ellington* (1968); *Citizens to Preserve Overton Park, Inc. v. Volpe* (1971); Strauss, "Revisiting *Overton Park*."

29. *Nashville I-40 Steering Committee v. Ellington*, 387 F.2d 179 (1967), 180–81.

30. *Nashville I-40 Steering Committee v. Ellington* (1967), 184–85.

31. *Nashville I-40 Steering Committee v. Ellington* (1967), 184–85.

32. *Nashville I-40 Steering Committee v. Ellington* (1967), 183–84.

33. *Nashville I-40 Steering Committee v. Ellington* (1967), 183–84.

34. Attempting to discern the Supreme Court's decision to deny certiorari in any case is generally a speculative endeavor. *Ellington* is no exception. While there is no conclusive archival evidence to support a clear hypothesis regarding the court's decision to turn down the *Ellington* appeal, the absence of such evidence has been discussed to great effect: See Campbell Haynes, "One Mile North," *Belmont Law Review* 8, no. 1 (2020): 35–39.

35. Henrik E. Stafseth, "Build, Baby, Build," *American Association of State Highway Officials Annual Conference* (American Association of State Highway Officials Annual Conference, 1969), 1.

36. Stafseth, "Build, Baby, Build," 2–4.

37. Stafseth, 2–4.

38. *Abbott Laboratories v. Gardner*, 387 U.S. 136 (1967).

39. Edwin Reis, "Route Alignment—Public Opposition or Litigation," *American Association of State Highway Officials Annual Conference* (American Association of State Highway Officials Annual Conference, 1969), 202–3; Pub. Law 79-404, 60 Stat. 237.

40. Schiller, "The Era of Deference," 399–441.

41. Reis, "Route Alignment—Public Opposition or Litigation," 202–3; Pub. Law 79-404, 60 Stat. 237.

42. Reis, "Route Alignment—Public Opposition or Litigation," 203.

43. Reis, 204–6; *Wilmington Park Authority*, 365 U.S. 715 (1960); *Jones v. Alfred H. Mayer Co.*, 392 U.S. 409 (1968).

44. Reis, "Route Alignment—Public Opposition or Litigation," 207. See also *D.C. Federation of Civic Associations v. Airis*, 391 F. 2d 478 (1968).

45. Reis, "Route Alignment—Public Opposition or Litigation," 207.

46. Reis, 208.

47. National Environmental Policy Act of 1969, Pub. Law 91-190, 83 Stat. 852. Reis, "Route Alignment—Public Opposition or Litigation," 206. See also *Scenic Hudson Preservation Conference v. Federal Power Commission*, 354 F. 2d 608 (2nd Cir. 1965), certiorari denied; *State of Washington v. Federal Power Commission*, 207 F. 2d 391 (9th Cir. 1953).

48. Edwin Reis, "Recent Environmental Cases as They Relate to the Location and Construction of Highways," *American Association of State Highway Officials Annual Conference* (American Association of State Highway Officials Annual Conference, 1970), 126–27 (emphasis mine).

49. *Citizens to Preserve Overton Park, Inc. v. Volpe*, 432 F.2d 1307 (1970); *Citizens to Preserve Overton Park, Inc. v. Volpe* (1971); Pub. Law 89-574, 80 Stat. 766; Pub. L 89-670 80, Stat. 931; Pub. Law 90-495, 82 Stat. 815; Thomas, "The Road to Overton Park," 433–58.

50. *Citizens to Preserve Overton Park, Inc. v. Volpe* (1970); *Citizens to Preserve Overton Park, Inc. v. Volpe* (1971).

51. *Citizens to Preserve Overton Park, Inc. v. Volpe* (1970), 1309–14.

52. "Mayoral Administration of Anthony J. Celebrezze," *Encyclopedia of Cleveland History*, Case Western Reserve University, accessed January 21, 2026. https://case.edu/ech/articles/m/mayoral-administration-anthony-j-celebrezze. The literature on the entwined histories of urban renewal, discrimination, racism, white flight, and poverty is vast. Studies particularly relevant to the context under analysis here include Cohen, *Saving America's Cities*; Kevin M. Kruse, *White Flight: Atlanta and the Making of Modern Conservatism* (Princeton University Press, 2005); and Thomas Sugrue, *The Origins of the Urban Crisis: Race and Inequality in Postwar Detroit* (Princeton University Press, 1996).

53. *Citizens to Preserve Overton Park, Inc. v. Volpe* (1970), 1318.

54. *Citizens to Preserve Overton Park, Inc. v. Volpe* (1970), 1317–18.

55. *Citizens to Preserve Overton Park, Inc. v. Volpe*, 432 F.2d 1307, 1317–18.

56. Reis, "Recent Environmental Cases as They Relate to the Location and Construction of Highways," 126–30.

57. Reis, 130–31.

58. Reis, 130; Pub. Law 89-670, 80 Stat. 931.

59. Lloyd Reeder, "Legal Requirements of Location Planning," *American Association of State Highway Officials Annual Conference* (American Association of State Highway Officials Annual Conference, 1971), 318–19; *Citizens to Preserve Overton Park, Inc. v. Volpe* (1971).

60. Reeder, "Legal Requirements of Location Planning," 318–19; *Citizens to Preserve Overton Park, Inc. v. Volpe* (1971).

61. Scholarship on the politics of race and environmental issues in American history covers a wide range of time periods and particular topics. Important and far-ranging analyses include Karl Jacoby, *Crimes Against Nature: Squatters, Poachers, Thieves, and the Hidden History of American Conservation* (University of California Press, 2014), and Dorceta E. Taylor, *The Rise of the American Conservation Movement: Power, Privilege, and Environmental Protection* (Duke University Press, 2016). A provocative general overview of the subject comes from legal scholar Jedediah Purdy, in "Environmentalism's Racist History," *New Yorker*, August 13, 2015.

62. Thomas, "The Road to Overton Park," 456–58; Strauss, "Revisiting Overton Park," 1259–60.

63. Pub. Law 79-404, 60 Stat. 237, § 706(2)(A); Schiller, "Enlarging the Administrative Polity," 1415.

64. Reeder, "Legal Requirements of Location Planning," 320; Deborah N. Archer, "'White Men's Roads Through Black Men's Homes': Advancing Racial Equity Through Highway Reconstruction," *Vanderbilt Law Review* 73, no. 5 (2020): 1313.

65. Reeder, "Legal Requirements of Location Planning," 320; Archer, "'White Men's Roads,'" 1313; Nathan D. B. Connolly, "The Strange Career of American Liberalism," in *Shaped by the State: Toward a New Political History of the Twentieth Century*, ed. Brent Cebul, Lily Geismer, and Mason B. Williams (University of Chicago Press, 2019), 64, 82; Connolly, *A World More Concrete*, 8, 183.

66. *Griggs v. Duke Power Co.*, 401 U.S. 424 (1971); *Swann v. Charlotte-Mecklenburg Board of Education*, 402 U.S. 1 (1971); *Milliken v. Bradley*, 418 U.S. 717 (1974); *Washington v. Davis*, 426 U.S. 229 (1976).

Chapter Six

1. Patrick Lohmann, "New York Selects 'Community Grid' Alternative for I-81 in Syracuse," *Post-Standard* (Syracuse), April 22, 2019, https://www.syracuse.com/news/2019/04/exclusive-new-york-selects-community-grid-alternative-for-i-81.html; Lanessa Owens-Chaplin, Johanna Miller, and Simon McCormack, "Building a Better Future: The Structural Racism Built into I-81, and How to Tear It Down," New York Civil Liberties Union, 2020, 4; Deborah Archer, *Dividing Lines: How Transportation Infrastructure Reinforces Racial Inequality* (W. W. Norton, 2025), 18–24, 70; "Syracuse: Redlining and Demographics," Segregation by Design, accessed January 30, 2026, https://www.segregationbydesign.com/syracuse/redlining; Jim Zarroli, "Why It's So Hard to Tear Down a Crumbling Highway Nearly Everyone Hates," *New York Times*, June 3, 2023, https://www.nytimes.com/2023/06/03/nyregion/syracuse-interstate-81.html; "Connecting the

Historic 15th Ward," City of Syracuse, accessed January 30, 2026, https://www.syr.gov/Departments/NBD/Initiatives/Connecting-the-Historic-15th-Ward.

2. Lohmann, "New York Selects"; Owens-Chaplin et al., "Building a Better Future," 4; Archer, *Dividing Lines*, 18–24; Segregation by Design, "Syracuse: Redlining and Demographics"; author's site visits and review of city maps.

3. Lohmann, "New York Selects." The community grid project later received $180 million in federal funding through the Inflation Reduction Act of 2022, and the state and city have continued to apply for federal grants. See Office of Governor Kathy Hochul, press release, March 13, 2024, https://www.governor.ny.gov/news/governor-hochul-and-syracuse-mayor-walsh-announce-i-81-viaduct-project-officially-receive-180; and Office of Mayor Ben Walsh, press release, February 20, 2023, https://www.syr.gov/News/City-News/City-2023/2023-02-20-Mayors-Office-News.

4. Michelle Breidenbach, "Syracuse Mayor Walsh to State: I-81 Community Grid Should Feel Like City Streets, Not a Highway," *Post-Standard* (Syracuse), October 22, 2021.

5. Maurice D. Lee and Nan Clements, "Mayor Rejects 'Threat,'" *Post-Standard* (Syracuse), September 13, 1963; "Sardino Aids Pickets, 97 Freed of Charges," *Post-Standard* (Syracuse), November 16, 1963.

6. See Congress for the New Urbanism, "Freeways Without Futures" (2023), https://www.cnu.org/highways-boulevards/freeways-without-futures/2023. In Joe Biden's presidential administration, Secretary of Transportation Pete Buttigieg also led novel efforts to redress infrastructural racism at a national scale, at the same time Syracuse was finalizing its community grid effort. See "Pete Buttigieg Launches $1 Billion Pilot to Build Racial Equity in America's Roads," NPR, June 30, 2022, https://www.npr.org/2022/06/30/1108852884/pete-buttigieg-launches-1b-pilot-to-build-racial-equity-in-americas-roads.

7. Jerome Cohn, "Urban Background to the Interstate Highway Program: The Planning and Politics of Highways in Syracuse, 1944–1960" (PhD diss., Syracuse University, 1978), 15.

8. "The United States in a New World," part 3, "The Domestic Economy," supplement to *Fortune*, December 1942, 4–5.

9. *Fortune*, "The United States in a New World," part 3, 7, 10, 14, 18–19.

10. "Syracuse Tackles Its Future," *Fortune*, May 1943, 121.

11. A group of political scientists, economists, and other academics from around the country analyzed *Fortune*'s impact on Syracuse—and Syracuse's postwar government more generally—throughout the 1950s. They published their findings in a lengthy volume in 1961 and revised it in 1968. See Roscoe Martin, Frank Munger, Jesse Burkhead, Guthrie Birkhead, Harold Herman, Herbert Kagi, et al., *Decisions in Syracuse* (Greenwood, 1968), 23. For the population statistics, see "Population of the 100 Largest Urban Places: 1940," United States Bureau of the Census, released online June 15, 1998, https://www2.census.gov/library/working-papers/1998/demographics/pop-twps0027/tab17.txt.

12. Cohn, "Urban Background," 17–18; *Fortune*, "Syracuse Tackles Its Future," 121.

13. Martin et al., *Decisions in Syracuse*, 52–53.

14. *Fortune*, "Syracuse Tackles Its Future," 120, 156.

15. *Fortune*, 158.

16. Archer, *Dividing Lines*, 70.

17. Jules Struck, "Forced to Live Here, Forced to Leave: The Twin Injustices of I-81 and the Demolition of the 15th Ward," *Post-Standard* (Syracuse), July 13, 2022, https://www.syracuse.com/news/2022/07/forced-to-live-here-forced-to-leave-the-twin-injustices-of-i-81-and-the-demolition-of-the-15th-ward.html.

18. "Visualizing 81," chap. 2, Syracuse University, accessed January 21, 2026, https://visualizing81.thenewshouse.com/chapter-2.html; Zarroli, "Why It's So Hard." Some insight into business life in the 15th Ward can be gained from *The Negro Motorist Green Book*—the famous guide, published from 1936 to 1966, for Black travelers seeking safe accommodation and service. The exact number of 15th Ward businesses listed in the *Green Book* fluctuated. The 1951 edition listed thirty-one, which was about average throughout the 1950s. See *The Negro Motorist Green Book: 1951 Railroad Edition; An International Travel Guide* (Victor Green, 1951).

19. Dunham is quoted in Zarroli, "Why It's So Hard."

20. See Walter Carroll, "Race Bias Bred Slums, Says 15th Ward Report," *Post-Standard* (Syracuse), February 24, 1954. On rent gouging, see Michael Kelly, "Running with the Land: Racial Capitalism, Restrictive Covenants, and the Pre-redlining Roots of the Private Real Estate Market in Syracuse, New York" (master's thesis, Syracuse University, 2021), 85.

21. "Visualizing 81," chap. 3, accessed January 21, 2026, https://visualizing81.thenewshouse.com/chapter-3.html; *Report of the Syracuse-Onondaga Post-War Planning Council* (1945), 65–67; Bureau of Public Roads, *General Location of the National System of Interstate Highways, Including All Additional Routes at Urban Areas Designated in September 1955* (US Government Printing Office, 1955), 62.

22. In the 1940s, Syracuse newspapers routinely described the 15th Ward as a "slum." See, for example, Richard Torrey, "City to Request Share of Fund, Mayor Asserts," *Post-Standard* (Syracuse), April 29, 1948.

23. For an overview of the connections between World War II and the Black freedom struggle, see Jacquelyn Dowd Hall, "The Long Civil Rights Movement and the Political Uses of the Past," *Journal of American History* 91, no. 4 (2005): 1233–63.

24. "Rogge Flays Costello, Education Board Ban," *Post-Standard* (Syracuse), March 9, 1949; "Street Orator Wins Court Stay," *Syracuse Herald-Journal*, March 9, 1949; "Syracuse Weather," *Post-Standard* (Syracuse), March 8, 1949.

25. Walter Wofford, "County ALP Organizes," *Syracuse Herald-Journal*, April 27, 1948; "ALP to Sponsor Three-Ward Rally," *Post-Standard* (Syracuse), August 15, 1949. On the general shift toward blaming officials and landlords for the 15th Ward's substandard conditions, see Kelly, "Running with the Land," 85, 111–13.

26. *Post-Standard* (Syracuse), "Rogge Flays Costello"; *Syracuse Herald-Journal*, "Street Orator Wins Court Stay." The Young Progressives of America were a leftist student group involved in the civil rights movement in the 1940s. They have been credited with playing a key role in Henry Wallace's 1948 presidential campaign, and were often tarred as communists by conservatives. See Thomas W. Devine, *Henry Wallace's 1948 Presidential Campaign and the Future of Postwar Liberalism* (University of North Carolina Press, 2013).

27. *Post-Standard* (Syracuse), "Rogge Flays Costello."

28. Roy S. Gutterman, "In 1949, Syracuse Radical Triggered Supreme Court Free-Speech Error That Echoes Today," *Post-Standard* (Syracuse), March 8, 2019, https://www.syracuse.com/opinion/2019/03/in-1949-syracuse-radical-triggered-supreme-court-free-speech-error-that-echoes-today-roy-s-gutterman.html; Douglas Martin, "Irving Feiner, 84, Central Figure in Constitutional Free-Speech Case, Is Dead," *New York Times*, February 2, 2009, https://www.nytimes.com/2009/02/03/nyregion/03feiner.html.

29. *Feiner v. New York*, 340 U.S. 315 (1951).

30. *Feiner v. New York*.

31. *Post-Standard* (Syracuse), "Rogge Flays Costello."

32. "Southwick Expects Second ALP Ban," *Post-Standard* (Syracuse), March 11, 1949.

33. *Feiner v. New York* (1951).

34. *Feiner v. New York* (1951); "Feiner Goes to Pen to Complete Term," *Syracuse Herald-Journal*, February 17, 1951.

35. "Breed Backs Conviction of Feiner," *Syracuse Herald-Journal*, August 7, 1949.

36. *Post-Standard* (Syracuse), May 7, 1949; "Hill Student Is Sentenced to 30 Days," *Syracuse Herald-Journal*, May 9, 1949.

37. "Feiner Goes to Jail," *Syracuse Post Standard*, February 16, 1951.

38. "100 Shiver at ALP Rally, City Officials Hit Again," *Post-Standard* (Syracuse), March 12, 1949.

39. "Campus Activism and Free Speech at Syracuse University."

40. *Feiner v. New York* (1951).

41. *Feiner v. New York* (1951).

42. *Feiner v. New York* (1951).

43. Author's review of city maps.

44. Archer, *Dividing Lines*, 26–27.

45. Denise Lynn, "Campus Activism and Free Speech at Syracuse University," *Black Perspectives*, March 19, 2020, https://www.aaihs.org/campus-activism-and-free-speech-at-syracuse-university/; Martin, "Irving Feiner"; Gutterman, "Syracuse Radical Triggered Supreme Court Free-Speech Error."

46. *Feiner v. New York* (1951).

47. Burton Halpern, "Morning Mail," *Post-Standard* (Syracuse), March 14, 1949.

48. John A. Williams, "Portrait of a City: Syracuse, the Old Home Town," *Holiday*, 1964; republished in *The Courier* 28, no. 1 (1993): 73, 81; "Walter Carroll's Last Story a Gripper," *Herald American* (Syracuse), March 19, 1989. On Republican Party dominance in Syracuse, see Martin et al., *Decisions in Syracuse*, 45–46.

49. Walter Carroll, "Negroes Denied Homes Except in Ghetto Sector," *Post-Standard* (Syracuse), January 27, 1954.

50. Walter Carroll, "Block-by-Block Cleanup Proposed in Slum Area," *Post-Standard* (Syracuse), January 22, 1954; Walter Carroll, "East Side Co-operative Formed in Effort to Solve Social Issues," *Post-Standard* (Syracuse), January 29, 1954.

51. Carroll, "Block-by-Block Cleanup"; Carroll, "East Side Co-operative."

52. Carroll, "Negroes Denied Homes."

53. Carroll, "Negroes Denied Homes," Walter Carroll, "Negro Problems in 15th Ward Seen as Outgrowth of Racial Prejudice," *Post-Standard* (Syracuse), January 28, 1954.

54. Carroll, "Negroes Denied Homes."

55. Walter Carroll, "Clinical Study Proposed to Point Way Toward Solving Ills of 15th Ward," *Post-Standard* (Syracuse), January 30, 1954; Carroll, "Negro Problems in 15th Ward."

56. Carroll, "Negro Problems in 15th Ward."

57. Carroll, "Clinical Study Proposed."

58. Carroll, "Clinical Study Proposed."

59. "Tanks Poised for Blitzkrieg in 15th Ward," *Post-Standard* (Syracuse), August 10, 1954; "Military Display in Watertown," *Post Standard* (Syracuse), May 17, 1953.

60. Walter Carroll, "47-Ton Punch Flattens 15th Ward Structures," *Post-Standard* (Syracuse), August 11, 1954.

61. Walter Carroll, "Landlord Discrimination Nearly Universal in City," *Post-Standard* (Syracuse), January 24, 1954.

62. "15th Ward Slums Cleared by Marines in 47-Ton Tank," *Post-Standard* (Syracuse), August 11, 1954; Joshua Allen, "Medium Tank Recovery Vehicle M74," Online Tank Museum, March 23, 2024, https://tanks-encyclopedia.com/coldwar/usa/m74-tank-recovery-vehicle/#google_vignette.

63. Carroll, "47-Ton Punch Flattens 15th Ward Structures."

64. *Post-Standard* (Syracuse), "Tanks Poised for Blitzkrieg in 15th Ward"; *Post-Standard* (Syracuse), "15th Ward Slums Cleared by Marines in 47-Ton Tank."

65. *Post-Standard* (Syracuse), "Tanks Poised for Blitzkrieg in 15th Ward."

66. "Probation Violator Sentenced to Pen," *Post-Standard* (Syracuse), September 17, 1954.

67. "A Conversation with James Baldwin," June 24, 1963, American Archive of Public Broadcasting, https://americanarchive.org/catalog/cpb-aacip_15-0v89g5gf5r.

68. Jeffrey Tucker, ed., *Conversations with John A. Williams* (University of Mississippi Press, 2018), 200.

69. John A. Williams, 1969 introduction to *Sissie* (1963; Doubleday, 1969), vii–viii, x.

70. Tucker, *Conversations with John A. Williams*, 108, 205.

71. Tucker, 204, 209.

72. William Grimes, "John A. Williams, 89, Dies; Underrated Novelist Wrote About Black Identity," *New York Times*, July 6, 2015, https://www.nytimes.com/2015/07/07/arts/john-a-williams-an-underrated-novelist-who-wrote-about-black-identity-dies-at-89.html; Williams, "Portrait of a City"; Tucker, *Conversations with John A. Williams*, 110, 135.

73. John A. Williams, *Sissie* (Farrar, Straus and Cudahy, 1963), 178.

74. David Boroff, "Blue Note for Bigotry," *Saturday Review of Literature*, March 30, 1963, 49.

75. Boroff, "Blue Note for Bigotry," 49.

76. Robert Phillips, introduction to the 1993 reprint of Williams, "Portrait of a City," 66–67 (see note 48, this chapter).

77. Eliot Fremont-Smith, "A Troubled Black, a Vivid Cowboy, and Two Vague Others," *New York Times*, November 13, 1967. Quoted in Williams's *New York Times* obituary years later; see Grimes, "John A. Williams."

78. John A. Williams, *The King God Didn't Save* (Pocket Books, 1971), 139. The radical perspective gained Williams a "dual reputation," *The New York Times* reported years later, "as a scathing critic of endemic racism in the United States and as a writer who, despite the constant comparisons to Ralph Ellison and James Baldwin, had been denied the credit due his talent." See Grimes, "John A. Williams."

79. "CORE to Attempt Stoppage," *Post-Standard* (Syracuse), September 13, 1963. After a few months, the city dropped charges against ninety-seven of the ninety-eight protestors; see *Post-Standard* (Syracuse), "Sardino Aids Pickets.

80. *Post-Standard* (Syracuse), "CORE to Attempt Stoppage"; "Stiffer Laws Seen Aid in Discrimination Fight," *Post-Standard* (Syracuse), October 3, 1963; *Post-Standard* (Syracuse), "Sardino Aids Pickets." See also Jay Fernandez, "Racism by Design: The Building of Interstate 81," American Civil Liberties Union, August 10, 2023, https://www.aclu.org/news/racial-justice/racism-by-design-the-building-of-interstate-81; and "Visualizing 81," chapters 1 and 3, https://visualizing81.thenewshouse.com/.

81. Williams, "Portrait of a City," 83.

82. Williams, 72, 83.

83. Williams, 72, 82.

84. Tucker, *Conversations with John A. Williams*, 73; Williams, "Portrait of a City," 70, 85–86.

85. Tucker, *Conversations with John A. Williams*, 73; Williams, "Portrait of a City," 70, 85–86.

86. Williams, "Portrait of a City," 71, 73.

87. Kenn Peters, "Rts. 81-690 Design a Costly Mishmash," *Herald-Journal* (Syracuse), January 30, 1972. See also "The I-81 Corridor Study," New York State Department of Transportation, 2013.

88. Dan Padovano, "$70 Million Plan to Ease I-81 Congestion," *Herald American* (Syracuse), June 22, 1980; Brittany Miller, "From Its Original Construction to Now, Here's Everything You Need to Know About I-81," *Syracuse Daily Orange*, November 17, 2022.

89. "I-690 Design Flaw Costs Taxpayers," *Herald American* (Syracuse), March 18, 1990.

90. Rachel Ramirez, "The New York Highway That Racism Built: 'It Does Nothing but Pollute,'" *The Guardian*, May 21, 2021, https://www.theguardian.com/us-news/2021/may/21/syracuse-new-york-highway-i81-viaduct-biden.

91. Alana Samuels, "How to Decimate a City," *The Atlantic*, November 20, 2015, https://www.theatlantic.com/business/archive/2015/11/syracuse-slums/416892/.

92. Paul Jargowsky, "The Architecture of Segregation," Century Foundation, August 9, 2015, 8, https://production-tcf.imgix.net/app/uploads/2015/08/07182514/Jargowsky_ArchitectureofSegregation-11.pdf.

93. New York State Department of Transportation, "The I-81 Corridor Study," 1, 11, 12, 21.

94. New York State Department of Transportation, 18, 22, 51.

95. "Let's Unite Syracuse: Replace I-81 with a Community Grid," editorial, *Post-Standard* (Syracuse), July 29, 2018, https://www.syracuse.com/opinion/2018/07/lets_unite_syracuse_replace_i-81_with_a_community_grid_editorial.html.

96. "Community Grid Is the Right Choice to Replace I-81," editorial, *Post-Standard* (Syracuse), April 23, 2019, https://www.syracuse.com/opinion/2019/04/community-grid-is-the-right-choice-to-replace-i-81-editorial.html.

97. Patrick Lohmann, "'There Are Literally Lives at Stake': St. Joe's Health Calls for Grid to Replace I-81," *Post-Standard* (Syracuse), August 9, 2018, https://www.syracuse.com/news/2018/08/st_joes_hospital_calls_for_community_grid_to_replace_interstate_81.html; Patrick Lohmann, "Gifford Foundation Board Endorses 'Community Grid' for Interstate 81," *Post-Standard* (Syracuse), August 1, 2018, https://www.syracuse.com/news/2018/08/gifford_foundation_board_endorses_community_grid_for_interstate_81.html; Patrick Lohmann, "Top Central NY Biz Group Endorses I-81 Grid—If There Are Extras," *Post-Standard* (Syracuse), February 22, 2019, https://www.syracuse.com/news/2019/02/top-central-ny-biz-group-endorses-i-81-grid-if-there-are-a-few-extras.html.

98. Lohmann, "New York Selects 'Community Grid.'"

99. Ramirez, "The New York Highway That Racism Built."

100. Fernandez, "Racism by Design."

101. Owens-Chaplin et al., "Building a Better Future," 14.

102. Owens-Chaplin et al., 14.

103. Owens-Chaplin et al., 9.

104. Owens-Chaplin et al., 10–11.

105. Owens-Chaplin et al., 12. The placement of a sewage treatment center and steam plant are representative of a broader phenomenon: Toxic development has often followed interstate highway construction, including landfills, power plants, power generation stations, and transportation hubs. All of this contributes to poor health outcomes in the vicinity of highway development. See Archer, *Dividing Lines*, 69–70.

106. Owens-Chaplin et al., "Building a Better Future," 17–18, 21–23.

107. Ramirez, "The New York Highway That Racism Built."

108. Quoted in Teri Weaver, "Plans for I-81 Pushing Forward amid Concerns from Some Syracuse-Area Businesses," May 1, 2013, https://www.syracuse.com/news/2013/05/plans_for_i-81_pushing_forward.html; see also Zarroli, "Why It's So Hard."

109. Zarroli, "Why It's So Hard"; Michelle Breidenbach, "NY State Is Free to Tear Down I-81 in Syracuse, Appeals Court Rules," *Post-Standard* (Syracuse), February 2, 2024, https://www.syracuse.com/news/2024/02/ny-state-is-free-to-tear-down-i-81-in-syracuse-appeals-court-rules.html.

110. *Renew 81 for All v. New York State Department of Transportation* (State Supreme Court, New York, 2023); amicus brief of the New York Civil Liberties Union in the case of *Renew 81 for All v. New York State Department of Transportation* (2022), https://www.nyclu.org/uploads/2022/12/007925_2022_renew_81_for_all_et_al_v_renew_81_for_all_et_al_affirmation_74.pdf.

111. *Renew 81 for All v. New York State Department of Transportation*, 224 A.D.3d 1273 (Supreme Court, Appellate Division, Fourth Department, New York, 2024); Michelle Breidenbach, "After a Mixed Decision in Court, NY Sticks to Plan to Remove I-81 Through Syracuse," *Post-Standard* (Syracuse), February 18, 2023, https://www.syracuse.com/news/2023/02/after-a-mixed-decision-in-court-ny-sticks-to-plan-to-remove-i-81-through-syracuse.html; Breidenbach, "NY State Is Free to Tear Down."

112. Michelle Breidenbach, "Opposition Group's Fight to Stop I-81 Project Is Finally Dead," *Post-Standard* (Syracuse), August 7, 2024, https://www.syracuse.com/news/2024/08/opposition-groups-fight-to-stop-i-81-project-is-finally-dead.html.

113. Zarroli, "Why It's So Hard."

114. New York State Department of Transportation, "The I-81 Corridor Study," 4–5; Valerie Wales, "Highway Removal in Syracuse and Buffalo: A Study in Interest Groups, Power, and Bureaucracy" (senior thesis, Princeton University, 2024), 36–37, 72.

115. Samantha Croston, "Community Grid Project: Could Increased Walkability Be a Path to Equity in Syracuse?" CNY Central, August 13, 2024, https://cnycentral.com/news/local/community-grid-project-could-increased-walkability-be-a-path-to-equity-in-syracuse; Scott Willis, "Syracuse Brings Its Community Grid Plan to All Parts of the City," WAER, August 9, 2024, https://www.waer.org/news/2024-08-09/syracuse-brings-its-community-grid-vision-plan-to-all-parts-of-the-city.

116. Michelle Breidenbach, "Syracuse Residents Lay Out Wishes, Demands for New Neighborhoods After I-81 Overpass Comes Down," *Post-Standard* (Syracuse), August 26, 2022, https://www.syracuse.com/news/2022/08/syracuse-residents-lay-out-wishes-demands-for-new-neighborhoods-after-the-i-81-overpass-comes-down.html.

117. Breidenbach, "Syracuse Residents Lay Out Wishes, Demands."

118. Michelle Breidenbach, "DOT Reveals Alternative to Controversial I-81 Roundabout," *Post-Standard* (Syracuse), February 18, 2022, https://www.syracuse.com/news/2022/02/ny-transportation-officials-reveal-alternative-to-controversial-i-81-roundabout-near-syracuse-school.html.

119. Lanessa Owens-Chaplin, "I-81 Plan Is Better for Neighbors—but NY Can Do More," *Post-Standard* (Syracuse), June 13, 2022, https://www.syracuse.com/opinion/2022/06/i-81-plan-is-better-for-neighbors-but-ny-can-do-more-guest-opinion-by-lanessa-owens-chaplin.html; Owens-Chaplin et al., "Building a Better Future," 14; Jeremy Boyer and Tim Knauss, "Rally Urges State to Keep I-81 Lands in Syracuse Out of City's Hands," *Post-Standard* (Syracuse), August 22, 2023,

https://www.syracuse.com/news/2023/08/rally-urges-new-york-to-keep-i-81-lands-in-syracuse-out-of-citys-hands.html.

120. Tim Knauss, "Black Leaders Angered and Mystified by Lawsuit That Halts Years of Work on I-81," *Post-Standard* (Syracuse), December 4, 2022, https://www.syracuse.com/news/2022/11/black-leaders-in-syracuse-angered-and-mystified-by-lawsuit-that-halts-years-of-work-on-i-81.html.

121. Zarroli, "Why It's So Hard."

122. Boyer and Knauss, "Rally Urges State to Keep I-81 Lands."

123. Breidenbach, "Syracuse Residents Lay Out Wishes, Demands."

124. Congress for the New Urbanism, "Freeways Without Futures."

125. Owens-Chaplin et al., "Building a Better Future," 31.

126. Owens-Chaplin et al., 31–33.

Conclusion

1. James Landis, *The Administrative Process* (Yale University Press, 1938), 70.

2. James C. Scott, *Seeing Like a State: How Certain Schemes to Improve the Human Condition Have Failed* (Yale University Press, 1998), 4.

3. The ARBA is quoted in Helen Leavitt, *Superhighway—Superhoax* (Doubleday, 1970), 116.

4. Deborah Archer, *Dividing Lines: How Transportation Infrastructure Reinforces Racial Inequality* (W. W. Norton, 2025), 24.

5. Archer, *Dividing Lines*, 158–60.

6. Deborah Archer and Joseph Schottenfeld, "Defending Home: Toward a Theory of Community Equity," draft, *University of Chicago Law Review*, November 14, 2025, 7–8, via SSRN, https://papers.ssrn.com/sol3/papers.cfm?abstract_id=4971493.

7. Paul Sabin, *Public Citizens: The Attack on Big Government and the Remaking of American Liberalism* (W. W. Norton, 2021), 152.

8. National Environmental Policy Act of 1969, Pub. Law 91-190, 83 Stat. 852.

9. Archer, "'White Men's Roads,'" 1314–15.

10. *Robertson v. Methow Valley Citizens*, 490 U.S. 332 (1989).

11. Archer, "'White Men's Roads,'" 1317.

12. The plaintiffs did not make claims under NEPA, but rather under New York's state-level counterpart, the State Environmental Quality Review Act, passed soon after NEPA, in 1975. But the same problems and shortcomings apply. See *Renew 81 for All v. New York Department of Transportation*, Supreme Court, State of New York, Docket No. 916 CA 23-00388 (2022).

13. Michelle Breidenback, "NY Is Free to Tear Down I-81 in Syracuse, Appeals Court Rules," *Post-Standard* (Syracuse), February 2, 2024, https://www.syracuse.com/news/2024/02/ny-state-is-free-to-tear-down-i-81-in-syracuse-appeals-court-rules.html.

14. Sabin, *Public Citizens*, 103–4.

15. Public hearings can, however, provide the basis of successful lawsuits that compel broadly desired agency action, an indirect but sometimes effective means of achieving citizen input. See Matthew Cortland and Karen Tani, "Reclaiming Notice and Comment," LPE Project, July 31, 2019, https://lpeproject.org/blog/reclaiming-notice-and-comment/.

16. Archer and Schottenfeld, "Defending Home," 37.

17. Archer, "'White Men's Roads,'" 1320, 1325–26.

18. Archer, 1321, 1326.

19. Archer, 1322.

20. Archer, 1327.

21. Archer, 1322, 1327–29.

22. Meg Jacobs, *Pocketbook Politics: Economic Citizenship in Twentieth-Century America* (Princeton University Press, 2007), 179–180; Sabeel Rahman and Hollie Russon Gilman, *Civic Power: Rebuilding American Democracy in an Era of Crisis* (Cambridge University Press, 2019), 138–40.

23. "CFPB Consumer Laws and Regulations," CFPB, manual V.2, October 2012.

24. Rahman and Gilman, *Civic Power*, 146.

25. Rahman and Gilman, 144–46. From its founding, the CFPB aimed to publish and update its enforcement actions continuously, on readily searchable webpages.

26. Rahman and Gilman, *Civic Power*, 144, 146–47.

27. Rahman and Gilman, 147.

28. For an overview, see Bijal Shah, "Executive (Agency) Administration," SSRN, June 12, 2024, https://papers.ssrn.com/sol3/papers.cfm?abstract_id=3353418##.

29. Cortland and Tani, "Reclaiming Notice and Comment."

30. John C. P. Goldberg, "On Being A Nuisance," *New York University Law Review* 99, no. 3 (2024): 926, via SSRN, https://papers.ssrn.com/sol3/papers.cfm?abstract_id=4648285.

31. Archer and Schottenfeld, "Defending Home," 39–40.

32. Jacobs, *Pocketbook Politics*, 230–31; Brent Cebul, *Illusions of Progress: Business, Poverty, and Liberalism in the American Century* (University of Pennsylvania Press, 2023), 151–54.

33. Rahman and Gilman, *Civic Power*, 148; the Supreme Court upheld the constitutionality of the CFPB's funding in *CFPB v. Community Financial Services Association of America, Ltd.*, 601 U.S. 416 (2024). In the early days of the second Trump administration, the White House attempted to shut down the CFPB through executive action. The US District Court in Washington blocked that action, but the case remains unresolved at the time of this writing. See Laurel Wamsley and Rafael Nam, "Federal Judge Blocks CFPB from Laying Off More Employees," NPR, February 14, 2025, https://www.npr.org/2025/02/14/nx-s1-5297870/cfpb-doge-trump-musk-federal-workforce.

34. Rahman and Gilman, *Civic Power*, 49.

35. Rahman and Gilman, 113, 134.

36. Reuel Schiller, "The Era of Deference: Courts, Expertise, and the Emergence of New Deal Administrative Law," *Michigan Law Review* 106, no. 3 (2007).

37. *Chevron U.S.A., Inc. v. Natural Resources Defense Council, Inc.*, 467 U.S. 837 (1984); Karen Tani, "Curation, Narration, Erasure: Power and Possibility at the U.S. Supreme Court," *Harvard Law Review* 138, no. 1 (2024): 68.

38. "*Chevron* Deference in the Court of Appeals," Congressional Research Service, June 8, 2023, https://crsreports.congress.gov/product/pdf/LSB/LSB10976.

39. *Loper Bright v. Raimondo*, 603 U.S. 369 (2024); Tani, "Curation, Narration, Erasure," 68–69; Cass Sunstein, "*Chevron* as Law," *Georgetown Law Journal* 107, no. 6 (2019): 1628, 1665, 1678.

40. Sabeel Rahman, "After *Chevron*: Political Economy and the Future of the Administrative State," LPE Project, July 23, 2024, https://lpeproject.org/blog/after-chevron-political-economy-and-the-future-of-the-administrative-state/.

41. Kate Aronoff, "This Is Why the Supreme Court Shouldn't Try to Do the EPA's Job," *New Republic*, June 28, 2024, https://newrepublic.com/article/183285/supreme-court-chevron-gorsuch-nitrous-oxide.

Index

Page numbers in italics refer to figures.